Technology and Power

David Kipnis

Technology and Power

Springer-Verlag
New York Berlin Heidelberg
London Paris Tokyo Hong Kong

David Kipnis
Department of Psychology
Temple University
Philadelphia, PA 19122 USA

With 2 Illustrations

Library of Congress Cataloging-in-Publication Data
Kipnis, David.
    Technology and power / David Kipnis.
      p. cm.
    ISBN-13:978-0-387-97082-0
    1. Technology—Social aspects.   2. Power (Social sciences)
  I. Title.
  T14.5.K57   1989
  303.48′3—dc20   89-11539

Printed on acid-free paper.

Typeset by Caliber Design Planning, Inc.

9 8 7 6 5 4 3 2 1

ISBN-13:978-0-387-97082-0   e-ISBN-13:978-1-4612-3294-0
DOI: 10.1007/978-1-4612-3294-0

For
Elliot and Andrew

Preface

There is a dark side to human nature that is nurtured by the control of power. In an earlier book, *The Powerholders*,[1] I described several psychological principles that appear to govern the behavior of people who control and use social power. In particular, I examined how the successful use of power transformed, for the worse, the values and behavior of the influencing agent.

My interest in the relation between technology and power grew out of reading David Howarth's *Tahiti: A Paradise Lost*,[2] a description of the almost causal ways in which Western technology was used by early explorers and traders to obliterate the Tahitian civilization. In reflecting on what happened in Tahiti, what struck me was the similarity in the behavior of these explorers and traders to the behavior of the husbands, wives, and businessmen, in positions of power, that I wrote about in my earlier book.

Technology and Power is concerned with the issue of how the added power provided by technology changes the behavior of people who control it. I describe these changes among managers at work, psychologists, physicians, and colonists. What unifies these disparate areas is the implacable logic of power. The seeming ease with which power promotes the derogation of those controlled by power provides, I believe, a needed perspective for viewing the many social problems generated by technology.

I would particularly like to thank my colleague, Stuart M. Schmidt. We have collaborated on several of the studies reported in this book. He has called to my attention literature concerned with the use of technology in industry, and, most importantly, we have spent pleasant hours discussing many of the issues of the book. I would like to thank equally Carolyn Tileston Johnson. She has taught me more than a bit about writing. Her helpful suggestions and comments are to be found throughout the book.

References

1. Kipnis, D. (1981). *The Powerholders*. Chicago: University of Chicago Press.
2. Howarth, D. (1983). *Tahiti: A Paradise Lost*. New York: Viking Press.

Philadelphia, Pennsylvania   David Kipnis

Contents

Part I   Introduction

The application of our intelligence to solve practical problems has given us power over events, things, and people. We begin by asking if people can use this power wisely.

CHAPTER 1

Technology and Human Needs

Psychologists have accounted for human behavior in many ways since the late 19th century. At first, behavior was seen as driven by instincts, then later by unconscious drives, and still later by needs and motives. Most recently, behavior has been explained by higher order mental processes involving intellectual activity, such as cognitions and expectations. What all of these explanations have in common is that they are based on social forces within the individual that cannot be seen or touched. These ghosts have such names as dissonance, id, and creativity, to name but a few. Each lives for a brief moment and then is forgotten as new explanations arise. Despite their being short-lived, psychology needs its ghosts, perhaps even more than the church needs the notion of a soul. How else are we to capture the richness and complexity of human beings? Surely not by behavior alone.

Critics point out, however, that psychology's explanations of behavior are not sufficient to account for human activities. That is, most psychologists attribute psychic turmoil and individual reactions to life to processes occurring within the individual – processes labeled neuroses, attributions, dissonance, or cognitions. The success of coping, then, becomes the index of the individual's psychological well-being. The events themselves are not studied as a means of understanding behavior.

As a psychologist, I do not discount the importance of internal psychological processes. Yet I also agree with critics that individual behavior is influenced far more by environmental events than traditional psychological thinking allows. Wars change our lives, technology changes the work we do, and economics changes our life-style. It makes a difference whether we live in a democracy or under totalitarian governments of the political left or right.

The assumption of this book is that technological changes in society interact with psychological processes to transform our social values and relations with others. These changes occur because of the particular predilection of humans for power and control. My goal is to account for these transformations. In particular this book will focus on changes in the social behavior of persons who gain power as a result of their control of technology.

The Technological Perspective

It is safe to assume that when humans first contemplated who they were, and what they wanted, they wanted more and more. Perhaps at first it was to obtain more food, or to find better shelter. Today, however, the catalog of human wants is endless. We want to cure sickness, eliminate injustice, and control our enemies. We want to live in luxury, make others do our bidding, reshape nature, or perhaps even live forever. If these wants were only satisfied we would be happy and learn to relax. Or would we?

There seems to be no end to what we want. But to get all of it, or even a part, takes power. Somehow we must devise techniques to overcome the resistance of nature and people. Thus the never-ending press of our desires seems to make the need for power and control a part of our intrinsic makeup. Only those who have given up wanting, as the Taoists tell us, are free from the need to exercise control. But for the rest of us, the more we want, the more we must focus on power.

Early on in history, physical force and simple tools were the principle techniques for getting what was wanted. And if these failed, there were always juju beads, spells, and incantations. Unfortunately, reliance on unseen demons, spirits, and gods to control events and people has been mostly a failure.

Only recently have people been able to exercise power over nature and each other. Not through beads and incantations, but through tools, machines, and applied science. Western technology has proved to be the way to make things happen. The magical spells of yesterday have been replaced by the marriage of applied mechanics and science. This combination allows us to achieve more uniform and predictable outcomes, with less effort, than could be produced by unassisted human effort. In sociologist Charles Perrow's terms, the use of technology reduces the amount of uncertainty in the attempt to solve practical problems.[1] By reducing uncertainty I mean increasing the probability that events will occur as wanted by an influencing agent.

Technology, then, represents the means by which humans exercise control over their physical and social worlds in order to achieve practical outcomes. José Ortega y Gasset defines technology more eloquently as the improvement brought about on nature by humans for the satisfaction of their necessities.[2] Necessities are imposed on humans by nature. Humans answer, according to Ortega y Gasset, by imposing changes on nature. Most animals survive by adapting to their environments. Humans survive by adapting the environment.

Thus, the history of human development is inevitably tied to the development of technology and its use. At its best, as David Dickson has observed, technology expands what people can do; it creates possibilities where none existed before.[3] Perhaps the major achievement of technology has been to reduce the amount of time and energy that we must devote to any given activity. Each new technological advance allows us to spend less effort in accomplishing what we want to do at work, at play, and in our social lives. The welcome relief from straining to push, pull, and shape objects by using our own physical strength surely must have been one of the initial goads to developing technology. Even today, new technologies readily gain acceptance if they promise relief from physical or mental effort.

Technologists can argue, then, with conviction, that applications of science have radically altered for the better each person's physical and social well-being. Developers of technology have even been compared to Prometheus defying Zeus, and bringing liberating knowledge to humans. Samuel Florman for instance, describes Prometheus as a revolutionary, as he says were Gutenberg, Watt, Edison, and Ford.[4] Technology, in short, is revolutionary since it overturns the conventional bonds of custom and habit. By these comparisons, Florman deftly turns the tables on political liberals who, he argues, usually oppose technology. It follows from his argument that hostility to technology is antirevolutionary, which is to say reactionary.

There seems no end to the social problems that technology can solve. Technological "fixes" are seen as the major way to solve societal problems. The former director of the Oak Ridge National Laboratory, Alvin M. Weinberg, recently described how technology has almost eliminated poverty[5]:

The traditional Marxian view of poverty regarded our economic ills as being primarily a question of maldistribution of goods. The Marxian solution was to eliminate profits, on the erroneous belief that this relatively small increment from the worker's paycheck kept him in poverty.

The Marxian view seems archaic in this age of mass production and automation. Advances in the technology of energy, of mass production, and of automation have created the affluent society. Technology has expanded our productive capacity so greatly . . . that there is more than enough to go around. Technology has provided the "fix" which enabled our capitalist society to achieve many of the aims of the Marxist social engineer without going through the social revolution that Marx saw as inevitable. Technology has converted the seemingly intractable social problem of widespread poverty into a relatively tractable one. (p. 31–32)

Not only does technology provide material benefits, but in its development there is also a potential for encouraging human growth, both psychological and spiritual. Theologian Norris W. Clarke, for example, sees technology as a liberating force, freeing people from servitude in primitive survival.[6] That is, technology frees human energies from their primitive state of almost total self-absorption in sheer brute physical labors as an essential condition for survival. By inventing more and more effective techniques for getting nature to work for them instead of against them, human beings free themselves progressively from spending their time in fulfilling their animal needs. The energy thus liberated can be diverted upward into various higher order needs. In the words of José Ortega y Gasset: "Technology provides men with the leisure to realize their true potential"[2] (p. 155). This, then, is the promise of technology: a material world of plenty and a spiritual world in which we have the leisure to realize our highest potentials and the freedom to contemplate God.

Yet in nagging counterpoint to this optimism about the beneficent world of plenty provided by unlimited power is the suspicion that technology has another face. There is also a world in which the freedom to choose and to control events has been subtly altered so that there is less choice and less control. This is the world of "megatechnics," to use Lewis Mumford's apt phrasing, in which technology concentrates power and reduces individual choice.[7]

It is this latter world that I will be discussing in this book. I shall argue that with each gain in control achieved through technology, the autonomous human extolled by Marx, Durkeim, and Neitzsche becomes less possible. Autonomy is lost because as technology makes things happen, there is less requirement that people use their inner energies and their skills to achieve what they want. Machines give people what they want with greater certainty, and with less effort. This results for many people in increased passivity, the use of fewer cognitive abilities, the lessening of dependence of people on their fellows, and the isolation of people from each other. In the home, for instance, games, reading, and social chitchat—entertainment that requires the active participation of family members —have been replaced by television and VCRs, entertainment that requires only passive monitoring of a screen. As we shall see, such changes have implications, not only for the individual, but also for the extent to which democratic processes can exist in society.

A Focus on the Powerholder

Of course these ideas about technology and power are not new. They have been described many times and in many ways since the early writings of Karl Marx.[8] But for the most part, the impact of technology has been examined in terms of its effects on people who are the targets of technology—workers, consumers, and citizens. All of these people have been shown to experience differential benefits and costs as their lives are altered by new inventions.

There remains, however, the nagging question of how added power provided by technology may change the lives of those who control the technology. With few exceptions, psychology and, to a lesser extent, sociology have not studied people who control power. For many reasons—historical, political, and economic— these disciplines have restricted their attention to the study of how people respond to power, that is, to studies of people's response to direct attempts to change their behavior, how they feel about these changes, and when they resist attempts to change their behavior.

Thus, we know much about how to change people's attitudes through various forms of persuasive arguments, how children respond to different kinds of child-rearing practices, how adults respond to various bases of power, and how behavior is shaped through different schedules of reinforcement, to name but a few of the many techniques developed by social science to understand and change the behavior of target persons.

What remains hidden from sight is the person who controls and manipulates the behavior of other people. In a previous book[9] I suggested that the ability to change people feeds back on the influencing agent in predictable, if not pleasant, ways. That is, as power is gained there are increasing tendencies to denigrate those who are the targets of influence. At the same time, the influencing agent's social values change: exploitive behaviors are justified, and there is an expressed desire by those who control power to increase social and psychological distance from persons whom they control.

To the extent that these kinds of changes occur, I would argue that there is much of social worth to be learned from the powerholder. One wonders, for instance, about how this individual seeks to control others. Does he or she use angry demands, cold logic, Machiavellian charm, or naked threats? We may want to know the circumstances that guide the choice of one or another of these various tactics. And of most importance, we want to know about the consequences to the powerholder of controlling other people's behavior.

These are questions that are hardly ever asked in the social sciences. As Harvard psychologist Herbert Kelman observed, "Psychologists prefer to study the powerless, rather than the powerful."[10] There appear to be many reasons for this.

First, people in positions of power can protect themselves from scrutiny. With few exceptions, information about how they use power, and how such control may have changed them is denied to investigators until long after the powerholders' deaths, when biographies appear. It is far easier to study people who are subjected to influence. Indeed, government and private granting agencies actively encourage the study of the powerless, that is, employees, children, the homeless, consumers, students, and the mentally ill.

Another reason for the study of persons who are targets of power is that the goal of most social scientists is to maintain and improve our social world. This frequently means devising ways of helping people whose behavior departs from the norms of society because of mental illnesses, poor parenting, lack of intelligence, poverty, and so on. These social concerns focus the psychologist's attention on ways to help change people.

For these reasons, psychological studies that have examined questions related to the control of power are few, and most of those are of the muckraking variety, or simply describe the economic status and life-styles of the elite. Yet there are many additional questions that can be asked. In earlier writings, I have reported empirical studies of the use of influence. These studies reveal that lawful psychological principles exist that describe how attitudes and behavior of influencing agents can be transformed by power. In this book I shall examine how this transformation is affected by the added leverage of technology.

Such information, I believe, is critically needed if we are to understand and control the many problems associated with technology. Most thoughtful discussions in this area either simply ignore the psychology of persons involved in decisions about technology or assume that these individuals are guided by universally shared values. A common assumption of most writers is that the harmful effects of technology can be reduced through careful planning and participatory decision making. Thus Paul Goodman can envision a humane medical technology based on small hospitals and neighborhood needs,[11] and union leader Mike Cooley can foresee worker control of decisions about how best to use technology.[12] Yet the reality is that hospitals grow larger and workers' jobs are made redundant in increasing numbers. What falsifies the reformers' visions, I believe, is that those who control technology do not share the humanistic values of the reformers. What seem to be rational solutions to outside observers are seen as irrelevant to those in positions of control.

In what follows, these issues will be discussed. I wish to show how ideas about the use of power provide a perspective for understanding decision-makers' seeming indifference to humanistic values. This book argues that technology not only changes the physical fabric of the world in which we live, but also alters psychological relations between people in unexpected, but predictable ways.

I will begin, however, by describing the psychological principles that appear to underlie the use of power and influence in daily lives. Included will be the many regularities that appear to guide people in the choice of influence strategies on a day-to-day basis. Why, for instance, does the same person yell, scream, and demand compliance in one instance, and plead and beg for favors in another?

Because the strategies we use to get our way not only change the behavior of those we influence but also inadvertently change ourselves as well, I shall next ask about the ways in which the use of power changes the powerholder. I have called these alterations "the metamorphic effects of power" to imply that the successful use of power transforms social relations between the more and less powerful. Suffice it to say here that it is the unilateral control of power that contributes to destructive conflicts and exploitive behaviors so often noted both in day-to-day life and in society at large.

The second part of this book applies the reasoning about the psychological consequences of using power to the control of technology. The general thesis is that the application of technology to solve social problems changes power relations, sometimes by design and at other times by accident. As a consequence, the behavior of those who control technology is altered in ways predicted by the metamorphic effects of power.

The concluding chapters will examine this contention by showing how technology transforms social relations between managers and their employees, physicians and their patients, social science practitioners and their clients, and colonialists and the native populations they control. Perhaps the major difference among these groups is in the strength of their attitudes toward the less powerful, rather than in their contents. I want to show by this analysis how the use of technology, rather than binding people in harmonious relations, tends to fragment social relations. This fragmentation occurs because technology tends to encourage the belief that persons who are subject to its influence are not in charge of their own behavior, and hence, can be treated with indifference.

Finally, a word of caution. This book is not meant as a critique of the social consequences of using technology, nor of the political and social factors that govern its development and use. I do not oppose technology, or believe that its use implies a Faustian pact with the devil. Rather, this book seeks to describe some heretofore unnoticed psychological consequences that result from technology's use. It is not my intent to show how technology is used to exploit the less powerful, nor to warn about the many threats to humanity that may result from its unthinking use. Thoughtful analyses of these issues are already available. What I see as lacking is an understanding of the ways in which the added leverage provided by technology alters the social values of those who use and control it.

References

1. Perrow, C. (1967). A framework for the comparative analysis of organizations. *American Sociological Review, 32,* 194–204.
2. Ortega y Gasset, José (1972). Thoughts on technology. In C. Mitcham & R. Mackey (Eds.), *Philosophy and technology,* (Chapter 23). New York: The Free Press.
3. Dickson, D. (1974). *Alternative technology and the politics of technical change.* Glasgow: Fontana/Collins.
4. Florman, S.C. (1981). *Blaming technology: The irrational search for scapegoats.* New York: St. Martin's Press.
5. Weinberg, A. (1966). Can technology replace social engineering? *University of Chicago Magazine, 59,* 6–10.
6. Clarke, N.W. (1972). Technology and man: A Christian view. In C. Mitcham & R. Mackey (Eds.), *Philosophy and technology,* (Chapter 10). New York: The Free Press.
7. Mumford, L. (1967). *The myth of the machine.* London: Secker & Warburg.
8. Easton, L., & Guddat, K.H. (Eds.) (1967). *Writings of the young Marx on philosophy and society.* Garden City, NY: Anchor Books.
9. Kipnis, D. (1981). *The Powerholders* (2nd ed.). Chicago: University of Chicago Press.
10. Kelman, H.C. (1972). Assignment of responsibility in the case of Lt. Calley. *Journal of Social Issues, 28,* 177–212.
11. Goodman, P. (1969). Can technology be humane? In C. Mitcham & R. Mackey (Eds.), *Philosophy and Technology* (pp. 335–354). New York: St. Martin's Press.
12. Cooley, M. (1980). *Architect or bee?* Boston: South End Press.

Part II   The Psychology of Power

In myth, power begins with the gods. They arrange human affairs as they choose. But in the social sciences, power begins with human beings' uncertain attempts to influence each other. And so we begin our account with a description of these uncertain attempts.

As we shall see, even these imprecise attempts to change people cause striking changes in the powerholder. Thus we can expect even greater effects when technology provides the means to control behavior.

Part II identifies the circumstances that guide individuals in their decisions to exercise influence. Why does the same individual, we ask, plead for compliance in one instance and threaten in the next? And more important, how does the successful use of influence change social relations between powerholders and those they influence. Does it make people humble, grateful, or arrogant to be able to "cause" behavior in others? The answer to these questions are needed if we are to understand the consequences of controlling people through technology.

CHAPTER 2

Tactics of Influence in Everyday Life

If you want you could give me money. Oh please darling, do that I beg you.
(Nora, *The Doll's House*, Act 2)

I had all the facts and figures ready before I made my suggestion to my boss.
(Manager)

I kept insisting that we do it my way. She finally caved in.
(Husband)

I sulked and looked unhappy.
(Lover)

I think its about time that you stopped thinking all these negative things about yourself.
(Psychotherapist)

Send out more horses, skirr the country round. Hang those that talk of fear. Give me mine armour.
(Macbeth, Act 5)

We know how to make automobiles go, but not people. By this I mean that technology allows control of many different aspects of our physical world, but as yet, little of our social world. As Perry London has noted, the technology of behavior control is still being developed.[1] Thus today, as throughout recorded history, a surprisingly large amount of our time is spent in activities that involve persuading people to do things. We try to convince people to accept our ideas, to do what we want, to leave us alone, and to provide us with material goods and services or with love and companionship, to name but a few of the many reasons for persuading others. Sometimes people do what we want, and other times they do not.

The statements at the beginning of this chapter are quotes describing how individuals attempt to convince others on a daily basis. With the exception of perhaps Macbeth's, the statements are not particularly subtle, Machiavellian, or statesmanlike examples of influence. Nevertheless, they do illustrate some of the many different ways in which people use words to influence others.

In some of the above instances, threats are used; in others, hurt emotions; and in still others, logic and nonemotional forms of persuasion. Why do these variations in influence tactics exist? Because many people have relied upon these or similar tactics at various times in their lives, the answer to the question may seem obvious. For instance, one common explanation is that the choice of tactics is based upon what "feels right" to use in that setting. A more pragmatic answer is that the choice of tactics is based upon the fact that its works, namely, that it persuades others.

A social psychological analysis of this question suggests, however, that less subjective answers can be provided by analyzing the personal and social context in which the influence act occurs. In this chapter I will examine verbal influence strategies at this social psychological level of analysis. I hope to show by this analysis that many of the psychological principles underlying the use of verbal tactics can also explain the use of technology as a strategy of control.

Why Bother to Influence Others?

I will begin this discussion with a brief look at why we need to influence other people. The answer, as social philosophers from Thomas Hobbes[2] in the 17th century, to R.M. Emerson[3] in the 20th century have observed, is that we are dependent on others to help us obtain material, immaterial, psychological, and spiritual things. Furthermore, there is no end to our dependence on others because there is no end to the things we want. The "dismal" observations of Thomas Hobbes—that people's motivations consist simply of endless streams of appetites—appear as true today as when he wrote *Leviathan*. When one need is gratified, new ones crowd in to take its place. We need other people to provide us with material goods, affection, information, and services, to name but a few of the many wants that require assistance from others. Only humanistic psychologists argue that as people mature psychologically, they become less and less dependent on others. However, the Maslowian ideal person is so rare as to be practically nonexistent.

Thus, the Hobbesian assumptions—that we are programmed by our desires, and by our dependence on other people to satisfy these desires—provide a framework for understanding the use of power. Furthermore, the more we want from others, and/or the more we believe they are unwilling to give us what we want, the more likely we are to invoke power and influence.

The relationship between the want and the inclination to influence others can be illustrated by recent studies of the use of influence by my colleague Stuart Schmidt and me.[4] We asked supervisors from a variety of industrial organizations how strongly they wanted to obtain various personal and organizational goals that were available to them at work, such as receiving more pay and better job assignments, receiving assistance on their jobs, having their bosses think well of them, having their bosses accept their ideas for new work projects or betters ways of doing things, and trying to get their bosses to work more effectively.

Some supervisors indicated that they had little or no interest in obtaining these various goals—they were satisfied with what they had already attained. Other supervisors indicated moderate interest in obtaining one or two of these outcomes, and still other supervisors very strongly wanted all of these goals. These latter supervisors tended to be slightly younger than the rest, and were certainly ambitious.

We also asked the supervisors how frequently they used various influence strategies when trying to convince their bosses. Some supervisors told us that they hardly spoke to their bosses, let alone tried to influence them. Some supervisors reported they used moderate amounts of communication and influence. Finally, there was a group of supervisors who continually "pestered" their bosses using a variety of influence tactics. They used flattery; they presented their bosses with logic, facts, and figures; they formed coalitions with other supervisors to bring pressure on their bosses; and at times they even demanded and insisted that their bosses do what they wanted. These were supervisors who refused to take "no" for an answer.

Not surprisingly, this latter group of supervisors were the ones who expressed the strongest wants in all areas of their work. They wanted their bosses to provide them with pay raises, better jobs, and the acceptance of their ideas, to name but a few of the many needs and desires that required their bosses' approval. They also used all available influence tactics to convince.

Wanting things that require the services of others, then, is the motivating force that encourages the use of power. The more we want, the more we are driven to influence others. To be free of this need to use power, be it at home, at work, or at play, requires that we give up wanting.

Control of Resources

Most theories of social influence state that the ability of the influencing agent to exert influence arises from the possession of scarce resources. Thus, wanting is only one part of the explanation for why we use power and influence. The second part of the explanation concerns the possession of resources that other people need. The more we have that other people want, the more dependent other people will be on our good will, and the more likely they are to do what we want. Furthermore, the more we see others as dependent on us, the more likely we are to try to influence them.

Thus, resources represent a powerholder's potential for exercising influence. William Gamson[5], for example, has whimsically called resources "influence in repose." The powerholder possesses something that the target person wants and cannot get elsewhere. The scarcer the commodity and the more it is valued, the greater the powerholder's potential for exercising influence. Thus, the powerholder may possess great wealth, be handsome or beautiful, be stronger than the target person, possess a gun; hold a position in an organization that allows him or her to influence others, or possess charisma, persuasive abilities, superior intelligence or expert knowledge. The reader may think of many other resources to which other people give weight, and that can, accordingly, be used as a base of power. The only two criteria for inclusion are that people (a) want what you have, and (b) cannot get it elsewhere. In short, the basis for power resides in scarcity and in dependency. This Hobbesian view suggests that the political left's platform of equal distribution of power, that is, "all power be given to the people," is perhaps less than realistic.

Why does the possession of appropriate resources increase the probabilities that the powerholder will take action? One answer from cognitive psychology is that the possession of appropriate resources needed by the target person raises the powerholder's expectations that he or she will gain compliance if influence is exercised. The individual who wants a new car is likely to approach the car salesman and make an offer only if he or she has money; supervisors will tell employees what to do because they control sanctions and a legitimate right to order employees; the therapist who wants to change a client's behavior will do so by invoking the specialized psychological knowledge that he or she possess; and the love-stricken suitor will gain the love of another by invoking charm and persuasive skills. In each of the above instances, expectations of successful influence are bolstered by the resources that each actor controls. Lacking the appropriate resources, each of the above actors must remain mute.

In short, the combination of wanting things that require the services of others and the possession of resources that are given weight by these others can be expected to produce influence, while the absence of either an aroused need or appropriate resources will lead to inaction.

Tactics of Influence

Tactics of influence refer to the actual means used by powerholders to change the behavior of other people. Tactics have been described in many different ways by social scientists. For instance, it is possible to use either words or nonverbal tactics such as smiling or glaring at someone else. One can even change people's environments, and so change their behavior.

Strength of Influence Tactics

Regardless of the particular modality that is used, I have found it useful to describe influence tactics in terms of the amount of freedom powerholders believe their use allows other people. Strong tactics of influence are those that are seen as not allowing people freedom to decide to comply without incurring severe costs. If a person complies to threats, for example, it is only natural for the threatener to believe that he or she forced the other person to comply. On the contrary, if people comply to tears and begging, the beggar will conclude that compliance occurred because people felt sorry for him or her. But the beggar is less likely to believe that tears forced compliance. Thus, one consequence of the use of strong tactics of influence is that it fosters the perception that the person being influenced is controlled by outside forces.

Since influence tactics are social acts, their meaning is subject to various interpretations depending upon the observer's vantage point. For example, a wife may say to her husband: "I wonder what we are going to do about the newspapers in the garage?" The husband, if asked, might assert that his wife is nagging him to clean up the garage. The wife, to the contrary, might protest that her remark was a tentative suggestion to her husband to consider the problem. Finally, an outside observer might report that the wife had not even tried to exercise influence, benign or aggressive.

Each perspective, then, provides a different interpretation of the influence act. In this book, the perspective of the powerholder will be used. Hence, when influence tactics are described as "strong" or "weak," I mean that the user of these tactics believes that their impact upon the person being influenced is strong or weak.

To illustrate, my students and I[6] have reported several investigations in which we examined how much control supervisors believed they exercised when they used influence tactics of various strength. Some tactics of influence involved the use of organizational sanctions, such as the offer of pay raises or threats of pay deductions in exchange for compliance. Other tactics consisted of managers simply discussing changes with employees, but letting the employees decide whether or not to comply. In these studies we found that the use of both promises of rewards and threats of punishment were seen by managers as forcing employees to comply; that is, these were strong tactics of influence. The supervisors had little doubt that their employees obeyed to get the rewards offered or to avoid the threatened punishments. On the other hand, when employees complied following the use of simple requests and discussion, managers attributed compliance to the voluntary choice of the employees.

Describing Verbal Influence Tactics

Suppose we wanted to develop a way of classifying animal life based upon the range and kind of tactics each animal uses when exercising influence. We would include such behaviors as vocalizations, staring, tail wagging, grooming, biting, and so on. No doubt this taxonomy would identify humans as displaying the widest range of tactics. Furthermore, this system would classify humans as developing new and powerful means of influence at a constantly accelerating pace. We are in a class by ourselves where power is concerned.

Within the field of psychology many methods of influencing behavior have been identified, ranging from making changes in an individual's environment to the use of nonverbal displays. Of these methods, the most frequently studied are the words that people use to persuade others. In particular, psychologists have devoted effort to classifying the many different verbal influence strategies that are used by people in ordinary conversations. The reason for this interest is that if we are to understand how influence is exercised, then it is necessary to be able to classify forms of influence into meaningful categories.

Empirical Studies of Influence Strategies

One method that gradually emerged during the 1970s was to ask people to describe the strategies they use in order to get their way. This empirical information was then grouped into logical categories of influence. For example, in 1969, Joseph Consentino and I asked supervisors to write essays describing an incident in which they corrected an employee's job performance.[7] The supervisors first described the problem presented by the employee and then described the ways in which they attempted to correct this problem. In telling how they corrected the subordinate's behavior, supervisors were describing the actual influence tactics that they used.

Over the last several years, similar essay reports have been used to describe the tactics used by dating couples, friends, husbands and wives, and even by business companies trying to influence each other.[8,9,10] Table 2.1 illustrates the many different categories of tactics that can be found by content analyzing people's descriptions of how they get their way. The incidents are based upon reports of employees describing how they tried to influence their superiors, peers, or subordinates.[11]

Clearly there are many different words that people at work use to "cause" behavior in others. Some rely on Machiavellian tactics of lying and pretending, others on flattery, some on sanctions, and still others on the use of logic, to name but a few of the strategies that people say they used. Of course, these many tactics raise the question of whether there is an overlap between the various categories of influence. For example, it is probable that the categories shown in Table 2.1 of Negative tactics, Demanded/ordered, and Persistence are different ways of using strong and controlling tactics.

To examine this question and move research forward, researchers have constructed questionnaires that contain each of the many tactics described by respondents in their written essays.[9,12,13] In answering these questionnaires, respondents describe how frequently they use each tactic to influence someone else (e.g., friend, lover, employee). These questionnaires are then statistically analyzed to determine the extent of overlap between tactics.

This procedure was used by me and my colleagues to study the kinds of tactics used by dating couples[8] and by people in organizations trying to influence their subordinates, peers, and superiors.[11] Because there were literally hundreds of individual tactics reported by people in these several settings, one of our first

TABLE 2.1. Classification of influence tactics used at work.[a]

Tactic	% Using	Tactic	% Using
Clandestine	8	Persistence	7
(Lied, flattered)		(Argued, repeated reminders)	
Negative tactics	8	Training	6
(Expressed anger)		(Showed how)	
Administrative	3	Rewards	2
(Filed report)		Direct requests	10
Self-presentation	5	Explain rationale	17
(Acted humble)		Begged/showed dependency	7
Demanded/ordered	7	Unclassified tactics	12
Formed coalitions	6		

[a] Taken from Kipnis, Schmidt, and Wilkinson, 1980.

objectives was to see if we could group individual tactics in terms of their similarity to each other. For this purpose, answers to the questionnaires were subject to factor analysis, a statistical technique that groups the tactics in terms of their underlying dimensions or similarity.

We found that three dimensions of influence best described the many different tactics used by couples when influencing each other, and that up to seven dimensions described the influence tactics used by organizational members. What was particularly exciting for us was the finding that the first three dimensions of influence used in affectionate relations and those used in business settings were very similar to each other, despite the fact that the words used in each setting were different. This suggests that there are common ways of influencing, regardless of the setting in which influence is used.

As shown in Table 2.2, respondents describing the use of influence in love relations and in business settings reported using strong and controlling tactics, weak

TABLE 2.2. Tactics illustrating the dimensions of influence used by couples and by managers in business organizations.

Couples	Managers
Use of strong tactics	
I get angry and demand that he/she do what I want.	I give orders in no uncertain terms.
As the first step in getting my way, I make him or her feel stupid and worthless.	I set a time deadline.
Use of weak tactics	
I act so nice he or she cannot refuse when I later ask for what I want.	I act very humble while making my request.
I show how much his/her stand hurts me by crying, sulking.	I make him/her feel important.
Use of rational tactics	
We talk about why we disagree.	I explain the reason for my request.
I offer to compromise, in which I give in a little, if he/she will give in a little.	I use logical arguments to convince.
Use of bargaining tactics	
	I offer an exchange (if you do this for me, I will do something for you).
Use of coalitions	
	I obtain the support of my co-workers to back me up.
Use of higher authority	
	I appeal to higher-ups.
Use of sanctions	
	I threaten an unsatisfactory performance evaluation.

and unobtrusive tactics, and tactics based on reason and logic. Somewhat similar findings have been reported by several researchers including Falbo[12] and Rausch, Barry, Hertel, and Swain.[14] These studies based their findings on influence tactics used by friends and lovers. The additional factors that emerged in the organizational analyses suggest that organizations give their members more ways of influencing others than are found in interpersonal settings.

When Are Stronger and Weaker Tactics Used?

Having identified three different ways of influencing that are used in affectionate relations and seven different ways of influencing that are used in organizational settings, and having devised questionnaires to measure these strategies, we can next ask when and for what reasons people choose to use each of these strategies. That is, why do people demand compliance in one instance, plead in another, and use reason and logic in a third?

The Setting

The choice of influence tactics is obviously constrained by the setting in which influence occurs. Custom and culture guide actors in their choice of tactics. Thus Latino cultures that emphasize machismo encourage men to use strong and controlling tactics and women to use weak tactics in their relations with each other. In contrast, Quaker societies rely on reason and logic to persuade.

Social relations between people also serve to guide people's choices of tactics. At a party of friends, for example, influence tactics differ from those that are used in formal business gatherings. This is because people are less concerned with impression management and giving offense in the former than in the latter setting. The climate or culture of a business will also shape choices of influence tactics. Stuart Schmidt and I have found that strong verbal tactics are likely to be used by managers to influence their employees when the firm is unionized, when organizational size is large, and when employees are unskilled.[15] Apparently in these circumstances, managers may encounter less cooperation. Hence, they use strong tactics rather than simple requests to overcome employee resistance.

As a general rule, when cultures and behavioral settings stress status and power, we are likely to find that people rely almost exclusively on the use of both strong and weak tactics to influence each others. In cultures and settings in which status and power differentials are less marked, people rely more often on nonemotional forms of influence, such as reason, to persuade. Settings that minimize power differentials between people, then, are associated with nonemotional forms of persuasion, and settings that emphasize inequality in power are associated with the use of emotional forms of persuasion that either demand or beg for compliance.

Balance of Power

Settings, then, restrain or encourage the kinds of tactics that will be chosen. Within any settings, however, the most important identifiable determinant of the choice of influence is the individual's resources, or power. Power enters into the selection of influence tactics in two ways. First, people who control resources that are valued by others, or who are perceived to be in positions of dominance, use a greater variety of influence tactics than do persons with less power. Second, people with power use strong tactics with greater frequency than those with less power. If the balance of power does not favor the individual, then weak tactics are used.

Variety of Influence Tactics

When the balance of power favors the influencing agent, and resistance is encountered, it is quite likely that the influencing agent will refuse to take "no" for an answer. Rather, he or she will be tempted to use all available means to overcome resistance. Thus, if one tactic does not succeed, added pressure will be placed on the individual by trying new tactics.

On the contrary, if the balance of power does not favor the influencing agent, it is very likely that he or she will give up trying to influence if resistance is encountered. This is because the powerless are unwilling to risk using tactics that are liable to offend the more powerful.

Thus, control of power gives influencing agents more options for getting their way. To illustrate, we asked managers both in this country and abroad[15] to complete questionnaires about the frequency with which they used each of the seven dimensions of influence shown in Table 2.2 when attempting to influence their employees. We also asked these managers to describe how frequently they used each of these dimensions to influence their own bosses. In all countries, managers reported that they used a greater range of tactics to influence their employees than their bosses. That is, these managers stated that they used strong tactics, weak tactics, bargaining, sanctions, and higher authority far more frequently with their employees. Only the strategy of reason was used to influence their bosses.

And so it goes at all levels in society; the control of power provides its possessor with flexibility to choose many different forms of influence tactics. Thus, parents can influence their children through rewards, threats, expert knowledge, their positions as "mommy" and "daddy," or even by changing their children's physical environments ("If you are good, we will give you a room of your own"). Children, on the contrary, have far fewer tactics they can realistically use to influence their parents, for example, asking, pleading, temper tantrums, and sulking.

Strong Tactics

In addition to being able to use a greater variety of influence tactics, people who have power are more likely to invoke strong and controlling tactics. While the relation between controlling greater power than the target of influence and the use of strong tactics is hardly surprising, it is found with great regularity both within organizations and general social relations. The link between controlling power and the strength of influence tactics that are used has been reported by social scientists among 6-year-old children attempting to influence younger children, peers, or teenagers;[16] among lovers attempting to influence each other;[8] among managers attempting to influence superiors or subordinates;[15] and among business organizations attempting to influence other businesses.[10] In all of these instances, when the balance of power favored the influencing agents, they used strong tactics to get their way. When the balance of power did not favor them, influencing agents chose weak tactics to overcome other people's resistance, such as acting nice, begging, and being polite. Thus 6-year-olds were reported by David Goldstein and his colleagues[16] to grab objects, yell, and demand, when trying to influence 4-year-olds, but to act polite and soft-spoken when trying to influence 18-year-olds.

The pervasiveness of this link between the control of power and the use of strong tactics has led to the speculation by Stuart Schmidt and me[15] that there is an "iron law of power," such that the greater the discrepancy in power between influencer and target, the greater the probability that the more powerful party will use strong tactics.

This does not mean that powerful persons demand and threaten as their first choice. Most people initially seek to exert influence through simple requests and the use of logic. Strong tactics are used only after the person being influenced refuses to comply, or the influencing agent anticipates refusal. When resistance is encountered, people with power escalate the pressure placed on target persons by shifting to the use of strong and directive tactics.

Goals of Influence

As I have pointed out earlier in this chapter, we are motivated to influence because we want something that requires the services of others. The kinds of services we want from others also guides us in our choice of tactics. Basically, people vary the tactics they use depending on whether or not they will personally benefit from influencing others. When we want favors, we rely on weak tactics to persuade. That is, people, more often than not, flatter, act humble, act more loving, or cry in order to obtain personal favors. When influence is not of direct benefit to the influencing agent, however, then stronger tactics are invoked.

To illustrate this process, a social psychologist, Samuel Fung[17] wrote vignettes that described various situations in which a persons was trying to influence others. In some of these vignettes the person was exercising influence to help out a friend; for example, to persuade the friend to study more often or to stop drinking. In other instances, the person was exercising influence to benefit him or herself; for example, to have the friend lend him money. For each vignette, respondents described the tactics of influence they would use.

Fung reported that weak tactics were chosen to obtain personal benefits and strong tactics were chosen when trying to benefit the other person. Very similar findings have been described in various other settings ranging from marriage to businesses. For instances at work, people state that they act humble and pleasant toward their bosses when they want, say, time off or better assignments. Yet the same people "insist" and "demand" when trying to cause changes that will improve the organization.

Thus people vary the strength of the tactics they invoke depending on what they want. The apparent explanation for this is that when an agent seeks personal favors from a target person, the target person gains power in the relationship, at least for the moment.

Expectations of Successful Influence

What I have stated so far would imply that people are involved in an elaborate calculus about power. Yet simple observation suggests that most people just act without thinking about how to influence. The choice of tactics remains, to use the psychologist Ellen Langer's terms,[18] a mindless act. In most instances we simply do not consciously evaluate our resources, the setting, and our goals before deciding about how to get our way. If you ask a friend, for example, why he acted so humble when he asked his boss for time off, he will tell you that yelling will only get him fired, or that the best way to catch flies is with honey. On the other hand, if you ask for an explanation as to why he threatened his little brother for noncompliance, he will tell you that he is "not going to take any noise from the little creep."

In short, the calculus about power is always there, but not explained in the language of the social scientist. In what ways, then, do the setting in which influence is exercised, the balance of power, and the agent's goals for exercising influence guide the choice of tactics?

What seems to happen is that all of the above enter the picture by raising or lowering the individual's optimism about his or her chances for success. If the setting is hostile, or one has little power, or one must rely on other's generosity for favors, then doubts creep in about one's ability to persuade.

As doubts—or in terms used by cognitive psychologists, expectations for successfully convincing others—change, so too does the choice of influence tactics. When optimism is high, most influencing agents rely on the use of simple requests, logic, and reason as the preferred persuasion tactics. As optimism or expectations of successfully influencing are reduced, then people depart from the use of simple requests, reason, and logical forms of persuasion. Rather, pressure for compliance is increased by broadening the range and emotional intensity of the tactics that are used.

Escalation takes two forms. When the influencing agent has power, but is pessimistic that others will comply, then escalation takes the form of strong and demanding tactics of influence. The powerful person just brushes aside resistance. If the influencing agent has no power in the situation, however, and is pessimistic that others will freely do what has been asked, then escalation takes the form of increased reliance on such tactics as begging, ingratiation, guile, and deceit. Not having the power to demand, our pessimistic agent must simply depend on the kindness of those he is seeking to influence.

Tactics of Escalation by Persons with and without Power

Let me illustrate how people with and without power use influence, after their requests have been refused, by citing findings from recent interviews with dating couples done by Irena Vilk and me.[19] In this research, we first determined the balance of power between each couple. We measured this balance by asking each respondent "who had the final say" when disagreements arose on such day-to-day issues as spending money, friends to see, physical relations, time spent together, and so on. Using this measure we could classify some individuals as dominant ("I have the final say"), others as submissive ("My partner has the final say"), and still others as sharing power equally. The latter men and women stated that they "decided together."

Next we asked these dating couples what influence tactics they typically used to change their partner's mind, and second, what tactics they used when their partners refused to do what was requested. They were asked to circle the tactics they used following refusal from among the following:

Strong Tactics

I get angry and demand that he/she give in.

I actively criticize his/her point of view as foolish, childish, absurd.

I keep repeating my point of view until he/she gives in.

Weak Tactics

I try to be especially sweet, pleasant, charming.

I show how much his/her stand hurts me by crying, sulking, looking unhappy.

I act so nice that he/she cannot refuse when I later ask for what I want.

Rational Tactics

I and my partner talk about why we don't agree.

I offer to compromise in which I give in a little and he/she gives in a little.

I back up my argument with facts and figures.

We found that men and women in the relationship who said they were dominant escalated the pressure placed on their partners by increasing their use of strong tactics and at the same time decreasing their use of rational tactics.

Their submissive partners also escalated the amount of pressure placed on their partners. However, escalation, in this instance, took the form of increasing the number of weak tactics they directed toward their partners. At the same time, these submissive respondents also decreased the use of rational tactics. Only men and women who shared power equally stated that they escalated the pressure placed on their partners to comply by increasing the use of rational tactics.

Thus persons with power overcome their dating partners' resistance by shouting and demanding. Their submissive partners, to the contrary, try to persuade their partners by increasing the frequency with which they cry and act nice. Only those who shared decision-making power equally sought to overcome their partners' refusals by increasing the use of reason and logic.

In short, the course of the influence process and the tactics that are used can be directly traced to dominance and power. Most people use reason and logic and simple requests most of the time until resistance occurs. At this point, beliefs about one's power clearly guide subsequent choice of tactics.

Gender Differences in the Use of Influence

Sex role stereotypes suggest that men should use strong and controlling tactics and women should use weak tactics to influence others. People who have looked at the use of influence in marriage and personal relations have reported findings that are consistent with this suggestion (see Howard, Blumstein, & Schwartz[20]). That is, women are found to use weak tactics and to avoid strong tactics when influencing their partners more often than their male counterparts.

Similar differences in the use of influence have also been observed in professional relations with clients and business associates. Thus, for example, Margaret Cooke and I[21] recently listened to tape recordings of male and female psychotherapists. We found that male therapists tried to influence their patients more frequently and interrupted their patients more often than female therapists. Both of these results are consistent with the stereotype of women as more passive and less assertive than men.

While few would dispute that there are differences in the use of influence by men and women, the reason for these differences is disputed. Some writers argue that differences in the exercise of influence are due to the early learning experiences of men and women. Feminist writers such as J.M. Bardwick,[22] M. Hennig and A. Jardin,[23] and Paula Johnson[24] propose that women are socialized to be compliant, submissive, emotional, indecisive, dependent, passive and noncompetitive. Hence, as adults they avoid the use of direct and controlling tactics of influence. This perspective has been extended to business settings. It is argued that women may not be effective as managers in settings where managerial roles require the use of assertiveness and strong tactics to influence others (see Hennig & Jardin[23]).

In contrast to this view is the argument that gender differences in the use of influence reflect differences in access to power, rather than to intrinsic differences between men and women. Social scientists such as Lisa Mainiero[25] and Roseabeth Kanter[26] argue that women at work are trapped in a cycle of powerlessness, which is the true cause of their use of weak influence tactics.

Based on similar reasoning, I would argue that gender differences in the use of influence can be attributed to the previously discussed factors of power, objectives, and expectations. In situations where men and women are relatively equal on these three variables, few gender differences in the use of influence tactics are likely to be found. Where these factors vary by gender, differences in tactics are to be found.

Support for this explanation has been reported in several investigations of gender differences in the use of influence at work. That is, when male and female managers are matched in terms of the resources they control,[25] their positions,[27] and their self-confidence[28] one finds few differences between men and women in their use of influence. Both men and women who lack confidence, or have little clout in the organization, are found to influence passively.

Even in personal relations, gender differences in the use of influence appear due to the above factors rather than to intrinsic gender differences. As I mentioned previously, we have found that the balance of power among married couples determines the mix of tactics that is used. That is, the dominant partner more frequently reports the use of strong tactics. It is noteworthy that among couples in which women are dominant (i.e., control decision-making power), women use strong tactics. Such findings support the notion that power, rather than gender, is the critical variable determining the strength of tactics that are chosen. Alice Eagly[29] has reported similar findings in a review of the relation between gender and social influence.

Other Studies of Influence Tactics

The information about verbal influence tactics and when they are used provides a reasonably new perspective for examining social behavior. To illustrate this contribution, this section examines how protagonists in drama seek to influence other characters in the play. I will show how protagonists' control of resources and their reasons for exercising influence both contribute to the frequency of the protagonists' attempts to influence, as well as to the strength of the tactics that are used.

Tactics Chosen by Protagonists in Drama

In plays, to the extent that plays mirror nature, people want things. They use words to influence. They control resources that are valued, or not, by others in the play. Surely then, we should find that protagonists in plays are guided in their use of influence tacctics by the same social circumstances as in real life.

We have asked how and why protagonists in each of four plays attempt to influence others. The protagonists are Lear, Macbeth, Hamlet, and Nora (in Ibsen's *The Doll's House*). In each play, I coded the number of times these four protagonists attempted to influence other people in the play.

In addition to coding the frequency of influence acts, each tactic was coded in terms of its strength. That is, tactics were coded as to whether the protagonist used stronger, weaker or rational tactics, illustrated as follows:

Stronger Tactics

"Sit down Torvald—we have a lot to talk over. Sit down, this will take time."
(Nora, Act 3)

"Kent, on thy life, I command thee now, no more."   (Lear, Act 1)

Weaker Tactics

"Pray do not mock me. I am a very foolish, fond old man."   (Lear, Act 5)

"Yes, take care of me Torvald, promise me that."   (Nora, Act 2)

Rational Tactics

"Think upon what hath chanced and at more time, the interim having weighed it, let us speak."   (Macbeth, Act 1)

"Speak the speech I pray you as I pronounce it to you."   (Hamlet, Act 3)

To illustrate the findings, Figure 2.1 shows the frequency with which Lear and Macbeth attempted to influence other people in the play. This information was obtained by first counting the number of times in each act that Lear and Macbeth attempted to influence other people and then dividing that number by the number of pages in the text in which these protagonists appeared on stage. Thus the information in Figure 2.1 represents influence acts per page.

What is surprising in Figure 2.1 is that Lear and Macbeth are attempting to exercise influence with relatively equal frequency throughout the play. Both are

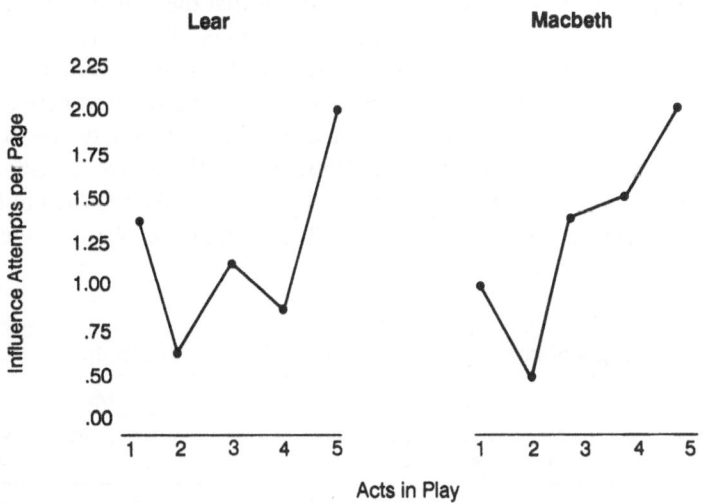

FIGURE 2.1. Frequency per page of influence attempts.

actively seeking to cause new behavior in others from the start to the end. In Act 5, both Lear and Macbeth are exercising approximately two attempts to influence per page. One would have thought that Lear would have stopped trying to get his way by this time, since his image in Act 5 is that of a feeble, dying old man. In fact, he is attempting to cause new behavior in others with greater frequency than in any of the preceding four acts of the play.

While Lear and Macbeth do not differ very much in the frequency of using influence, they do differ in terms of when they use strong tactics. Table 2.3 shows the percent of influence tactics that were coded as strong, weak, or rational in each act. The base for these percentages was the total number of tactics used in each act.

What we find in Table 2.3 is that Lear's tactics become increasingly weaker over the five acts of the play and Macbeth's become increasingly stronger. These changes in the strength of influence tactics are consistent with what has been said earlier, that the strength of influence is guided by the resources controlled by the powerholder. Since Lear has given up his major base of power, his throne, he must attempt to influence others with weak tactics. Macbeth, who has gained a kingdom as his base of power, can employ strong means of influence.

This comparison allows us empirically and systematically to supplement the usual intuitive critical assertion that Lear is a play about the loss of power whereas Macbeth is a play about the struggle to gain power. The loss and gain in power refer more precisely to the changes in the strength of influence tactics that are used rather than to any change in the frequency of their attempts to influence others. These attempts remained the same for both. Both protagonists have many wants that require the services of other people. What they have to do, however, is adjust the strength of their requests to be consistent with the resources they control.

I have already mentioned the Hobbesian thesis that the motivation for exercising power originates in our dependence upon others. The more we want from others, the greater the frequency of our attempts to influence others.

The use of influence tactics by Hamlet illustrates the relation between wanting things from others and frequency of influence. A major theme of Hamlet is his attempts to change his own behavior, rather than anyone else's, and so avenge the death of his father. This inward focus is directly reflected in Hamlet's attempt to influence others in the play. Throughout the play, Hamlet makes

TABLE 2.3. Strength of influence attempts used by Lear and Macbeth.

	Lear					Macbeth				
Tactic	Act 1	Act 2	Act 3	Act 4	Act 5	Act 1	Act 2	Act 3	Act 4	Act 5
Strong	64%	57%	13%	14%	0%	33%	36%	44%	75%	77%
Weak	16%	38%	25%	79%	100%	33%	36%	9%	19%	4%
Rational	23%	5%	63%	7%	0%	33%	27%	47%	6%	19%

far fewer attempts to influence other people than does Lear or Macbeth. The latter protagonists make three times as many attempts to influence other people as Hamlet does.

In *The Doll's House*, Nora's attempts to influence her husband also illustrate the twin ideas that our dependency on others and what we want from them motivate the use of influence. During the first two acts of the play, Nora expresses many needs for personal favors that can only be satisfied by influencing her husband. These range from the need to receive love and affection to the need for money to pay the blackmailer, Nils Grogstand.

As a result of these many needs, Nora is continually seeking to influence her husband. During the first two acts, Nora directs toward her husband almost three influence attempts per page, far more than either Lear, Macbeth, or Hamlet. In the final act, however, Nora's attempts to influence her husband drop to an average of less than one-half an attempt per page. This drop reflects the fact that Nora is no longer dependent on her husband. She has rejected the love and support she formerly sought. As a result, she no longer has any reason to influence him.

The influence tactics that Nora uses are also consistent with what has been said about the relation between the choice of strong tactics and the control of resources. In Act 3, Nora no longer wants anything from her husband, since she has rejected his love and support. To shift the balance of power even further in Nora's direction, her husband now realizes his dependence on Nora. Nora then controls the balance of power in the relationship. This shift in power is accompanied not only by a decrease in the frequency of Nora's attempts to influence, as I described above, but also by a shift from the use of weak tactics in Acts 1 and 2 to the use of strong tactics in Act 3. In Acts 1 and 2, over 80% of Nora's tactics directed toward her husband were classified as weak tactics and 6% were classified as strong. In Act 3, 23% were classified as weak and 53% as strong.

To summarize, dramas illustrates very well, I believe, how the choice of influence tactics is guided by predictable events. People are neither Caligulas nor Timid Souls. Rather, their style of influence is guided by life's circumstances. The protagonists in the above plays were also guided by life's circumstances and used influence tactics in ways that are consistent with the use of influence in real life. That is, the frequency with which they attempted to influence, and the strength of the tactics they used, were guided by their resources and what they wanted from others.

Technology, Dependency, and Power

In the preceding sections, primary consideration was given to the conditions that determine whether or not strong tactics will be used. It was noted that the strength of tactics chosen by influencing agents is guided by such factors as the setting, what influencing agents want (i.e., their goals), and their expectations of being able to influence successfully. Settings in which inequality is the norm, reasons for influencing are impersonal, or the agent anticipates resistance from the target encourage the selection of strong tactics. Above all, we have seen that one-sided power relationships encourage the use of strong tactics. In this section, we consider how technology feeds this one-sided power relationship by providing influencing agents with additional power.

Earlier in this chapter, I noted that the elements of a definition of power include the possession of scarce resources and the dependency of people on these resources. From this perspective, I suggest that technology alters the balance of power by altering people's dependencies. There are at least three ways in which this occur.

Technology Reduces Dependency on Other People

Perhaps the most often discussed way in which technology shifts power relations is by taking over the creative aspects of human work. Each new invention makes obsolete some particular combinations of individual skills and talents. The skills involved in commercial weaving, printing, sewing by hand, and machining tools and metals, have each in their turn been replaced by technical processes.

One consequence for employees of losing control over the cognitive and skill components of work is that they lose power, and management gains power. John French and Burt Raven's[30] theory of social power provides a very reasonable perspective for understanding these shifts in power. In this theory, knowledge and expertise are resources that other people give weight. We are dependent on experts to help us solve our own problems. At work, the engineer, accountant, machinist, and computer programmer are valued because they can contribute to the organization's functioning. High salaries, status, and respect are commanded by employees who possess the expertise needed by others. In nonwork settings, the skilled artist, chef, raconteur, and athlete are valued for their competencies and abilities to carry out activities that most people cannot. Having knowledge, skills, and abilities needed by others allows experts to exercise influence. Being an expert increases both an individual's feelings of competency and other people's dependency on that individual's knowledge and talents. People listen to what such individuals have to say. Who seriously listens to what janitors have to say? Even with the best of intents, people in positions of authority are unable to emplower persons doing "mindless" work.

The formula, then, is simple. having knowledge that others need equals power. When people no longer depend on our knowledge, however, we lose power. Harry Braverman's[31] influential critique of management–labor relations is based on this formula. In his book, *Labor and Monopoly Capital*, Braverman describes management's use of technology as a deliberate strategy to reduce labor's bargaining power. He argues that as work is de-skilled, management is able to dictate terms of employment, or even to eliminate the need for a labor force.

Needless to say, Braverman's analysis of technology as a deliberate strategy used by capital to wrest power from labor has been vigorously debated (e.g., Form,[32] Littler & Salaman,[33]). One objection has been that in socialist countries, as in capitalist one, employees doing routinized work are found to have less influence than employees doing skilled work. Sociologist William Faunce,[34] for instance, documents that the common requirements of industrialization in both Marxist and capitalist countries make for a high degree of similarity in the worker's ability to exercise power. Such findings implicate technology rather than capitalism as the underlying cause of workers' loss of power.

While the explanation for the motives involved in using routinization may be contentious, few would disagree with the conclusion that the de-skilling of work reduces the power of the worker. As we no longer need to depend on the skills of the craftsman, the craftsman become less able to influence us.

In short, the first, and most often remarked upon, way in which technology alters power relations is by reducing the individual's need to display competency. In turn, this reduces our dependence on the individual's skills, and hence the individual's ability to exercise influence over us, leaving us free to treat him or her as we will.

Technology Creates New Dependencies

Another way in which technology shifts power relations is by increasing our dependence on its products. In turn, our dependence provides those who control the technology with added power. Thus, technology not only takes power away from some people, but it also empowers others. For example, in the 1950s and 1960s, relatively few people knew how to program and operate computers. As a result, computer technicians had considerable bargaining power with management. They were paid large salaries and provided with many perks and benefits in order not to leave. In this instance, technology empowered computer specialists by increasing management's dependence on their professional knowledge.

A further illustration of the way in which technology empowers can be found by examining changes in the status of the medical profession over the last 100 years. Paul Starr and others[35,36] have described how physicians in the 18th and 19th centuries had very little real power in American society. This was because the physician's understanding of disease, and his ability to control it, were limited. The physician was more often caricatured as a quack and county fair medicine man than as a healer and man of science.

Today, however, few would describe physicians as quacks. The power of physicians in our society has increased enormously over the last 75 years. this has occurred because medical technology has found real cures for diseases and developed new operative techniques to replace and heal our bodies. As a consequence of these technological advances, people now perceive physicians as experts, capable of healing sickness. As technology has increased the certainty of healing, people's dependency on physicians has shown a corresponding increase. Few patients today would refuse to carry out the physician's orders, no matter how difficult or threatening. In short, as technology finds new ways to solve important social problems, those who control the new technology gain in power as people seek the benefits of these new solutions.

Technology Directly Controls Behavior

In their visions of future worlds, Orwell and Huxley describe people enslaved by drugs, weapons, remote forms of coercion, and mind-altering devices. Through these means "masters" are seen as gaining power and exercising control over the "numbed masses".

In fact, the amount of effort devoted to developing such forms of control are minuscule. Only in the area of military technology do we find consistent attempts to develop techniques for controlling people, and these attempts involve weapons, not mind-control techniques. Technology's main agendas are concerned with reducing the nagging uncertainties of daily effort and increasing the scope of human activities. For the most part, control of people is an accidental by-product of attempts to realize these goals. Yet despite its accidental nature, the introduction of many new technologies, as I have suggested, changes people's dependency on each other, and consequently empowers those who direct and use the technology.

This final section discusses one important way in which technology does provide "masters" with additional control by providing techniques for monitoring the behavior of dependent persons. It is now possible to monitor the health status of patients in hospitals, the location of prisoners under house arrest, and even the military and industrial activities of entire nations by means of electronic recording devices, remote signaling beepers, and satellites, to name but a few of the methods now available for observing events through remote means. Customers in department stores can be placed under television surveillance, or monitored by electronic scanning devices. At work, computer technology can be used to monitor the pace and quality of employee's performance, at times without their knowledge or consent.

Psychologists such as L. Strickland[37] and A. Kruglanski and M. Cohen[38] have independently shown that the ability to place individuals under surveillance has psychological consequences for both those who do the watching and for those who are watched. Social psychological studies find that people who are allowed to monitor the behavior of others, while remaining unobserved themselves, eventually distrust and depersonalize relations with those they have placed under surveillance. The consequence for the individual of being watched on the other hand, is to reduce behavioral freedom. That is, the behavior of persons who are watched is less varied, and conforms closely to the rules and standards of those who are watching.

Clark Molstad[39] described how this process of remote monitoring worked on a production line in a brewery.

The production lines are in continuous operation and any pause in the process at any point causes the line to start backing up. When it backs up, automatic light sensors will shut the line down, red lights will flash, and warning horns will sound. These warnings will immediately claim the attention of supervisors. Everyone will start looking to see where the trouble is. The lights and horns will identify the problem in a matter of seconds. If the problem is mechanical failure, machinists will fix the trouble. However, if the problem is caused by a machine operator who is not keeping up with the machine, the operator will be chastised and embarrassed. He or she will be watched until the production line again flows smoothly, or will be demoted to a less desireable job. (p. 4)

Molstad observed that because of this production technology, supervisors do not have to spend time observing and controlling production workers. The technology watches for them by sounding the alarm at the first hint of trouble.

In her account of the use of computer-based technology in industries ranging from paper and pulp plants to international banks, Shoshana Zuboff[40] provides rich detail about the ability of computers to monitor employee behavior. While the uses of the computers varied from business to business, what was common in all settings was that employees had to log onto the computers. In turn, the computer could then electronically record information about many different aspects of the employee's performance, including mistakes, speed of work, and even the nature of messages transmitted from one work area to another. Zuboff labels the ability of computers to monitor this information the *panoptic power of information technology.* By panoptic power, Zuboff means the use of computer information by management to observe employee behavior, at the discretion of management, and without the employees knowing exactly when they are being observed.

Zuboff reports interviews with plant management that suggest how the added control provided by surveillance depersonalized social relations between management and employees. One manager who relied on computer surveillance (The Overview System) reported:

If I didn't have the Overview system, I would walk around and talk to people more. I would digress, like asking someone about their family. I would be more interested in what people are thinking about. (p. 330)

Another manager stated:

It's beautiful now. I can track my people's work. All I have to do is type the craftsman's initials in (the computer) and see how he is progressing and see what his total load was. What is his productivity. Before we had to judge people more on hearsay. Now we have it in black and white. (p. 331)

The information provided by surveillance is readily translated into direct control of behavior; thus, a plant manager reports:

We have disciplined and terminated people based on information from the Overview System. It provided information on incidents which showed that the individual was not performing to basic requirements. By recording what happens to all on a five second basis, we can see what was done, what should have been done, and what was not done. Since you know the people that were there, you know what they did or did not perform. (p. 336)

As I mentioned, surveillance by authority serves to limit the behavioral freedom of those being watched. Conformity to expected behavior increases simply because employees are never sure when their behavior is being monitored. An employee Zuboff interviewed who had just been promoted to foreman after many years as a craftsworker, stated the consequences of being watched this way:

I hated it. It was too close. I could no longer hide anything. Management could monitor me hour by hour, and that was kind of scary. (p. 352)

In one of the surveyed companies described by Zuboff, computers were used so that managers and executives could hold conferences. The advantages of computer conferencing was that people in different locales could communicate, while avoiding the formality of written correspondence. As people got used to the idea of talking to each via computers, informal chitchat increased. Thus one set of conferences were held as Computer Coffee Breaks in which conference members soon expressed free wheeling and often derogatory comments about the company. Unfortunately, the conferees believed that their messages to each other were confidential. This was not true. All computer communications were recorded and reviewed by top management. When it became known that management could monitor these freewheeling comments, participants opted for anticipatory conformity and after a few months, the Computer Coffee Breaks were dropped as too risky.

In sum, the ways we seek to control others are linked to the norms of society and its various settings on the one hand, and to the control of power on the other. We have seen in this chapter that power originates from many sources, both personal and institutional. Ultimately, however, power resides in people's dependencies. We gain in power as people depend on us for, say, information, affection, and care. Our ability to exercise power changes as scientific and technical advances alter the extent people depend on the resources we control. I have presented evidence that as power increases, influencing agents are drawn to the use of strong and controlling tactics. We appear to move with little thought from concerns about respecting the rights of persons we wish to influence, to a desire to restrict their rights, as our power increases, and people appear reluctant to do what we want. But there is a price to pay for the use of strong tactics. We may get our way, but control changes us as well as those we influence. In Chapter 3, I will examine the nature of these changes and then, in Part III, apply these insights to the control of technology.

References

1. London, P. (1969). *Behavior Control*. New York: Harper & Row.
2. Hobbes, T. (1968). *Leviathan*. England: Penguin Books.
3. Emerson, R.M. (1962). Power-dependence relations. *American Sociological Review, 27*, 282–298.
4. Schmidt, S., & Kipnis, D. (1982). Managers' pursuit of individual and organizational goals. *Human Relations, 10*, 781–794.
5. Gamson, W.A. (1968). *Power and discontent*. Homewood, Il: The Dorsey Press.
6. Kipnis, D. (1981). *The powerholders* (2nd ed.) Chicago: University of Chicago Press.
7. Kipnis, D., & Consentino, J. (1969). Use of leadership powers in industry. *Journal of Applied Psychology, 53*, 460–466.

8. Kipnis, D. (1984). The use of power in organizations and in interpersonal settings. In S. Oskamp (Ed.), *Applied social psychology annual 5*. Beverly Hills, CA: Sage.

9. Cody, M.J., & McLaughlin, M.L. (1980). A multidimensional scaling of three sets of compliance-gaining strategies. *Communication Quarterly, 3*, 34–36.

10. Wilkinson, I., & Kipnis, D. (1978). Interfirm use of power. *Journal of Applied Psychology, 63*, 315–320.

11. Kipnis, D., Schmidt, S., & Wilkinson, I. (1980). Intraorganizational influence tactics. *Journal of Applied Psychology, 65*, 440–452.

12. Falbo, T. (1977). Multidimensional scaling of power strategies. *Journal of Personality and Social Psychology, 35*, 537–547.

13. Falbo, T., & Peplau, L.A. (1980). Power strategies in intimate relations. *Journal of Personality and Social Psychology, 38*, 618–628.

14. Rausch, H.L., Barry, W.A. Hertel, R.K., & Swain, M.A. (1974). *Communication, Conflict, and Marriage*. San Francisco: Josey-Bass.

15. Kipnis, D., & Schmidt, S. (1983). An influence perspective on bargaining. In M. Bazerman & R. Lewicki (Eds.), *Negotiating in organizations*. Beverly Hills, CA: Sage.

16. Goldstein, D., Miller, K., Griffin, M,. & Hasher, L. (1981). *Patterns of gestures and verbal communication in a tutorial task*. Unpublished manuscript. Department of Psychology, Temple University, Philadelphia.

17. Fung, S. (1987). *The social influence process: With particular emphasis on the effects of power, relationship, purpose, and strategy*. Unpublished doctoral dissertation, Temple University, Philadelphia.

18. Langer, E.J., Blank, A., & Chanowitz, B. (1978). The mindlessness of ostensibly thoughtful action. *Journal of Personality and Social Psychology, 36*, 635–642.

19. Kipnis, D., & Vilk, I. (1987). *The escalation of influence tactics among dating couples*. Unpublished manuscript, Temple University, Philadelphia.

20. Howard, J.A., Blumstein, P., & Schwartz, P. (1986). Sex, power, and influence tactics in intimate relationships. *Journal of Personality and social Psychology, 51*, 102–109.

21. Cooke, M., & Kipnis, D. (1986). Influence tactics in psychotherapy. *Journal of Consulting and Clinical Psychology, 54*, 22–26.

22. Bardwick, J. (1971). *Psychology of women*. New York: Harper and Row.

23. Hennig, M., & Jardim. A.(1977). *The managerial women*. New York: Pocket Books.

24. Johnson, P. (1976). Women and power: Towards a theory of effectiveness. *Journal of Social Issues, 32*, 99–109.

25. Mainiero, L. (1986). Coping with powerlessness: The relationship of gender and job dependency to empowerment-strategy usage. *Administrative Science Quarterly, 31*, 633–653.

26. Kanter, R.M. (1977). *Men and women of the corporation*. New York: Basic Books.

27. Izraeli, D.N. (1982). *Gender differences in self-reported influence among union officers*. Paper presented at the annual meeting of the American Psychological Association, Washington, D.C.

28. Instone, D., Major, B., & Bunker, B.B. (1983). Gender, self-confidence, and social influence strategies. *Journal of Personality and Social Psychology, 44*, 322–333.

29. Eagly, A.H. (1983). Gender and social influence. *American Psychologist, 38*, 971–981.

30. French, J.R.P., & Raven, B. (1959). the bases of social power. In D. Cartwright (Ed.), *Studies in social power*. (pp. 150–167) Ann Arbor, MI: University of Michigan, Institute of Social Research.

31. Braverman, H. (1974). *Labor and monopoly capital*. New York: Monthly Review Press.
32. Form, W. (1980). Resolving ideological issues on the division of labor. In H.M. Blalock, Jr. (Ed.), *Sociological theory and research: a critical approach* (pp. 140–161). New York: The Free Press.
33. Littler, C.R., & Salaman, G. (1982). Bravermania and beyond: Recent theories of the labour process. *Sociology, 16,* 251–269.
34. Faunce, W.A. (1981). *Problems of an industrial society*. New York: McGraw-Hill.
35. Starr, P. (1982). *The social transformation of american medicine*. New York: Basic Books.
36. Shorter, E. (1985). *Bedside manners: The troubled history of doctors and patients*. New York: Simon and Schuster.
37. Strickland, L. (1958). Surveillance and trust. *Journal of Personality, 26,* 201–215.
38. Kruglanski, A., & Cohen, M. (1973). Attributed freedom and personal causation. *Journal of Personality and Social Psychology, 26,* 245–250.
39. Molstad, C. (1988). Control strategies used by industrial brewery workers. *Human Organization, Fall,* 1–12.
40. Zuboff, S. (1988). *In the Age of the Smart Machine*. New York: Basic Books.

CHAPTER 3

The Metamorphic Effects of Power

We change in many ways and for many reasons. Disasters and setbacks cause us to become wary of others. Love causes us to become happy and generous. We change because of natural events, such as growing older. We change because we are dissatisfied with ourselves or because others express dissatisfaction.

Sometimes we change our behavior to conform to religious teachings, or to a physician's warnings. In these instances, we may stop drinking and carousing. At other times we change to conform to images from popular culture or from advertising messages. Perhaps we do not like our appearance and so seek to transform our looks through diet, exercise, and cosmetics. Or perhaps we feel that we are too shy, insufficiently sensitive, or not assertive enough. For these psychological and social deficits, we turn to mental healings, charismatic leaders, meditation, diets, education, gurus, cults, and sometimes even religion. All offer the promise of transformation.

One agent of change that hardly anybody uses is power. Only the weak and the dispossessed recognize that the control of power may be the way to improve their lives. But even in these instances, power is sought as a means of improving material well-being. Hardly anyone advocates control of power as a means of changing one's personality or spirituality. And yet it is precisely in these areas that the control of power does produce changes in individuals.

In Chapter 2 I described the kinds of verbal tactics that people use to influence each other and identified the circumstances under which people choose one or another of these tactics. Here, I turn to the vexing problem of how and why the use of such tactics may change the influencing agent. I have called these changes the *metamorphic* effects of power. By this term I want to suggest that the use of power transforms social relations between the more and less powerful in predictable, if not particularly pleasant, ways.

For the most part, discussions of metamorphic effects has been the province of political scientists, authors, and biographers of great leaders. The field of psychology has been relatively silent about the links between the control of power and behavior. With the exception of the writings of psychologists such as Phillip Zimbardo[1], little has been written about the circumstances under which power may cause changes in the normal person.

My goal then in this chapter is to give an account of why good people who achieve power do "dirty works." I will argue that the continual exercise of successful influence changes the influencing agent's views of others, regardless of whether the actors involved are managers influencing their employees, married couples influencing each other, or great political leaders who are responsible for the well-being of entire nations. The transformations are the same in each instance. They are brought about, in my opinion, as the result of ordinary psychological processes that relate to how we perceive and interpret events.

What kinds of transformations may be expected as a result of being able to cause behavior in others? History, literature, and psychological research provide parallel accounts in which the control of other people's behavior and thoughts encourages the belief that those we control are less worthy than ourselves. Being less worthy, people can be used and exploited in ways that would not be acceptable among persons of equal status. Thus, power, control, devaluation, and exploitation of others are intertwined, perhaps inevitably.

Early Writings

The existence of metamorphic effects has been recognized almost from the earliest writings about the use of power. The Greek dramatists were particularly sensitive to the fate of persons who were at the high tide of their power and status. In the plays of Sophocles, for instance, the reader is confronted with the image of great and powerful rulers transformed by their prior successes so that they are filled with a sense of their own worth and importance—with hubris—impatient of the advice of others and unwilling to listen to opinions that disagree with their own. Yet in the end, they are destroyed by events which they discover, to their anguish, they cannot control. Oedipus is destroyed soon after the crowds say (and he believes) that "he is almost like a God." King Creon, at the height of his political and military power, is brought down as a result of his unjust and unfeeling belief in the infallibility of his judgments. Sophocles warns us never to be envious of the powerful until we see the nature of their ends. Too often, arrogance, bred of power, finally causes its own defeat.

So, too, in many of the plays of Shakespeare are the themes of power, arrogance, cruelty, and eventually destructive consequences told again and again. King Lear destroys his family, his friends, and finally himself with his unthinking demands for flattery and obsequious behavior.

A persistent image in literature and political science is that of an individual, virtuous and innocent at the time he assumes power, soon transformed by his own success into, at worst, a tyrant, or at best an insensitive and immoral person. In this century we have seen revolutionary leaders who, as young men and women were motivated by the selfless idea of bringing freedom and equality to all, in the end were transformed into the most inhuman of leaders, devoid of feeling for those whom they set out to free.

These transformations are not restricted to persons exercising political influence. The unilateral control of power has similar consequences at all levels of society. Thus, in marriage, the husband or wife who exercises complete control over the affairs of the marriage is found to express contempt for the submissive spouse. In their classic account of life in the Indian village of Karimpur during the 1920s, the Wisers[2] observed the same transformations occurring among agents of absentee landlords – agents who, once appointed to their positions, exploited their fellow villagers. As the Wisers state; "If you were to take one of the most harmless men in the village and put him in the watchman's place, he would be a rascal within six months and (soon) . . . a hardened tyrant" (p. 113).

Power Corrupts

The well-known observation of Lord Acton that "power tends to corrupt and absolute power corrupts absolutely" captures the idea of the kinds of transformations that may occur as a consequence of being able to cause behavior in others.

Before presenting what I believe are the psychological bases for these transformations, I want to briefly describe the various ways in which power has been seen as corrupting individuals. This discussion allows an understanding of the areas focused upon by the metamorphic model. It is based in part on a schema by political scientist H. Lasswell for describing the effects of power (see Rogow & Lasswell[3]).

Pursuit of Power as a Life Goal

The first meaning assigned to the observation "power corrupts" refers to the belief that those who gain power tend to value it above all other values and restlessly pursue additional power throughout their lives. The corrupting influence of power, in this view, is that power becomes an end in itself and replaces the Christian values of love, charity, compassion for the weak, and the like. The urge to be "number one" becomes the exclusive preoccupation of the powerholder. When faced with the choice between giving up power or maintaining it by less than moral or legal methods, those with a taste for power choose the second option.

Power as a Means to an End

A second meaning attached to the idea that power corrupts refers to the idea that the individual uses power illegally to enrich him or herself. In these instances, one usually finds the powerholders hold a position in which they have access to institutional resources, or where they can exchange favors of their offices for bribes. In Philadelphia, where I live, an instance of this was recently revealed when several municipal judges were discovered to be accepting payments from officials of the Roofers Union in exchange for giving lenient sentences to roofers. Thus, power corrupts in this second usage because it tempts individuals to deviate from the formal duties of their public roles in order to enrich themselves.[4]

Power, Morality, and Self-Concepts

The corrupting effects of power can also refer to changes in powerholders' self-evaluations and changes in their moral values. Sorokin and Lundin,[5] in a review of the behavior and attributes of individuals controlling political and economic power, state that persons holding power develop an exalted and vain view of their own worth. Sorokin and Lundin further suggest that powerful persons evolve new codes of ethics that serve to justify their use of power.

These changes appear to involve at least two related processes. First, powerholders may find themselves the recipients of flattery and well-wishing from those who are anxious to keep in their good graces. It is quite common for powerholders to receive positive feedback from the less powerful, both true and false, concerning their own worth.

Second, powerholders frequently find their ideas and opinions are readily agreed with. Common sense tells followers that it may be costly to disagree continually with persons who can affect their fate. This public agreement and compliance may lead powerholders to believe that their ideas are superior to others' ideas. I believe that as a result of this continued deference and flattery, it is almost inevitable that powerholders should come to believe that they are something special.

In his recent biography of Fidel Castro, Tad Szulc[6] describes how this aging leader has become impatient with those who offer even the mildest of criticisms. Castro, Szulc relates, does not listen to advisers who suggest that his ideas about socialistic control are wrong; he views the failures of his policies as due to traitors and worse.

Because they believe they are cut a notch above the ordinary person, many powerholders believe that they are not subject to ordinary rules. Frequently, powerful persons evolve new codes of ethics that serve to justify their use of power. That is, there is one set of ethics for exchanges with persons of equal worth, and a different set for persons who are less powerful than themselves. Throughout history, we find that a special divinity is assumed to surround the powerful so that they are excused from gross acts such as murder, theft, terrorism, and intimidation. Machiavelli, for instance, wrote that it is necessary for a prince to learn how not to be good. That is, the prince must do those things, whether good or evil, that will perpetuate his own power. One has but to read the history of the murder of the Jews by the Germans during World War II to understand how power obliterates morality.

The corruption of power, then, in this third usage focuses on how powerholders' self-concepts are changed by continued flattery, compliance, and false feedback, as well as how morality is altered to justify the use of power.

Power and Evaluation of Others

A fourth meaning assigned to the idea that power corrupts focuses on how powerholders devalue the worth of the less powerful and act to increase social distance from them. Here then, in my opinion, is one of the more destructive psychological consequences of one-sided power relationships—the transformation that occurs in how the more powerful see the less powerful. From individuals with both strengths and weaknesses, the less powerful become objects of manipulation with a lesser claim on human rights than claimed by the powerholder. In Martin Buber's terms, it is the transformation of one person's perceptions of another from "thou" to "it," from individual to object.

Considering these tendencies to devalue others and to maintain psychological distance, many writers believe that the control of power precludes the possibility of harmonious interpersonal relations. According to Sampson,[7] inequity in power inevitably produces dominance and manipulation, and precludes the possibility of establishing truly loving relations. "At a minimum," according to Sampson, "the deference and compliance shown by the less powerful is seen as a sign of weakness, if not servility."

Now I would like to point out that there is a mystery here that should not be ignored. In theory, it is perfectly reasonable to expect that we should like those we influence, since they are doing what we requested. The fact that we instead dislike them requires explanation.

After all, to dislike a cooperative and compliant neighbor requires an acceptable reason, unless we assume that we hate for the sake of hating. Most psychological explanations of attraction between persons suggests that we dislike persons when we are competing for scarce goods, when bargaining fails so that we cannot reach a mutually advantageous agreement, when we differ in values and attitudes from others, and when another person's entry into a goal region prevents us from entering. But to express contempt for a person who is compliant and does what we want makes little sense. It is counter to the idea that we like people and events that provide us with positive reinforcements.

Nevertheless, there are explanations as to why compliance can cause the influencing agent to dislike rather than like the target of influence. The first is that it is easier to influence others if psychological distance is maintained and emotional involvement is kept to a minimum. This is especially true if powerholders believe that it is likely that they will order the less powerful to carry out behaviors that are distasteful. To the extent that powerholders feel sympathy for the position of the less powerful, they may not want to issue orders. It is psychologically comfortable to assume that the person being influenced is not as worthy as oneself. Powerholders can, then, in good conscience make the target person do things that they would not willingly do themselves.

A second reason why devaluation of the target person occurs is more subtle, less dramatic, and yet of greater interest in terms of its implications for understanding the use of power. This has to do with the possibility that the very act of influencing causes devaluation of the target person. This possibility appears to be particularly likely when the powerholder invokes controlling or strong means of influence.

I suggest that when strong tactics of influence are used, the powerholder believes that the target person is not in control of his or her own behavior. Rather, the behavior of the person being influenced is seen as caused by the power-holder's orders and suggestions. In essence, the locus of control is seen to reside in the powerholder, who attributes causality for change to himself. "There is no need to give him credit," says the powerholder, "He simply followed my orders, step by step."

If what has been said is correct, then the frequency of devaluation of target persons may be very pervasive in situations where there is an imbalance in power. Rather than being limited to master–slave relations, it is possible that devaluations occur in any unbalanced power relationship—student–teacher, dominant wife–subordinate husband—where the powerholder demands compliance and the target person obeys. The very act of compliance under these circumstances may diminish the worth of anything achieved by the target person in the opinion of the powerholder. This is because the individual is seen as not in charge of his or her own behavior. The loss of autonomy produced by power becomes the focus on which outside observers, the powerholders, and even those who are controlled base their evaluations.

In the following sections, I will amplify the above explanations to account for the ways in which the control of power transforms the powerholder. First, to summarize, I have pointed out that the corrupting influence of power can refer to the fact that (a) persons acquire a "taste for power" and restlessly pursue more power as an end in itself; (b) access to power tempts individuals to use institutional resources illegally as a means of enriching themselves; (c) with the control of power, persons are provided with false feedback concerning their own worth, and develop new values designed to justify their use of power; and (d) at the same time they devalue the worth of the less powerful and prefer to avoid close social contact with them.

Before proceeding further in the discussion of the psychological consequences of controlling people, I want to examine the concept of individual autonomy in more detail. By this discussion I want to show why evaluations of people are so closely tied to beliefs about the extent that they can control their own behavior.

Autonomy and Control

The question of how much freedom and how much control is best for the ordinary person is continually debated by social philosophers, theologians, politicians, and sometimes by psychologists and people in general. Are people happier when they are told what to do, or when they can decide for themselves? Is society better off when people are told what to do? Within families, we find the same ambivalent attitudes toward autonomy. Mothers and fathers want their children to "fit in" and "get along" with their schoolmates. They are told not "to rock the boat" or "make waves." On the other hand, being a conformist and "refusing to stick up for one's ideas" is considered equally unacceptable.

Questions about autonomy have been answered in varied ways depending on the individual's political stance, position in life, philosophy, and assumptions about the nature of human beings. It is not unusual for people in power to argue that autonomy should be unequally distributed—those of equal power to themselves should be autonomous and those with less power should learn to submit to outside forces. At this extreme is the argument, perhaps best expressed by Thomas Hobbes in *Leviathan*, that people and society are better off when strict controls are enforced. People left to decide for themselves are inevitably driven to competition and strife in their unending pursuit for power.

This position is repeated with dramatic intensity in *The Brothers Karamazov*. In this novel, Dostoyevski describes an apocalyptic dialogue between the Grand Inquisitor and Jesus Christ, who has returned to 15th-century Spain. The Grand Inquisitor vigorously defends the church's need to establish order and eliminate independent thought. He argues that people are happier when they are relieved of the burden of making decisions for themselves. To allow people to decide when to pray, where to work, how to behave, and what to think leads to chaos. The strong church, the strong state, and the strong leader provide the guidance and the security that people need. Thus control preserves society and the welfare of people in it.

I think it is fair to say that whenever unilateral control exists for long periods of time, those in charge justify control by saying that people are happier freed of the burden of autonomy and/or that they are not yet ready to decide for themselves. In modern political life we find echoes of the Grand Inquisitor when leaders of the political right and the political left assure the world that they will soon install democracy in their countries—but not yet. For the moment, their citizens are better off led by an elite. Ordinary people are not ready to accept the responsibilities of democracy and to decide for themselves. Arguments against autonomy can also be found in Marxist analyses of individualism. They see it as simply a ruling class device that encourages the poor to blame themselves, rather than society, for their lack of achievement.[8]

People themselves are frequently found willing to give up autonomy. Japanese culture, for example, stresses the importance of conformity to group norms as proper behavior. In Western societies, people give up autonomy when they are troubled, threatened, or in psychic or physical pain. Then they willingly submit to individuals who seemingly have the power to take care of them. For instance, psychologists J. Thibaut and H. Walker[9] have observed that people in conflict accept the decisions of a court-appointed arbitrator rather than attempt to resolve the conflict themselves. Willing submission is also found among persons who are sick or psychologically ill. These individuals eagerly seek out the physician or counselor who will tell them what to do in order to relieve their distress. In short, persons who believe they are not in control of their lives, for whatever reason, willingly submit to control. Moreover, as the Grand Inquisitor argued, such people are happier now that others of greater ability have taken charge of their lives.

Individual autonomy can also be seen as problematic because the autonomous individual is selfish and pursues individual interests at others' expense. Feminist psychologists Rachel Hare-Mustin and Jeanne Marecek[10] have suggested, for example, that women's identity may be rooted in relatedness, connection with others, and emotional dependence. They argue that the autonomous individual seeks prominence and is drawn into conflict with others. The result is a loss of community and concern for one's fellows. Thus autonomy is a male value that may require the sacrifice of feminine values of community, sharing, and capacity for intimacy. For women in therapy it may be inappropriate for the therapist to advocate autonomy. Conformity to others, argue Hare-Mustin and Marecek, may, in fact, be psychologically healthier than attempts to achieve individual prominence.

There is, however, another answer to the question of how much freedom and how much control is best for the ordinary person. A contrary answer to these questions is given by 19th-century philosophers such as Marx, Durkheim, and Nietzsche. These writers argue that the essence of humanity is the ability of people to be their own autonomous source of action. Sociologist Ernest Becker[11] in describing human growth and development writes:

Man is free when he enjoys a rich participation in a broad panorama of life experiences; when he dwells in an expansive present that responds to his own energies. In one sense the primitive (man) is the most free because he is free to make his integral powers felt in the world; because he senses that the control over his life stems in large part from within himself, and that therefore he can and should assume responsibility for it. (p. 247)

With but few exceptions, modern psychological theory also describes self-determination as the mark of the healthy individual. Psychological theories concerned with human development—as proposed, for example, by Sigmund Freud, Eric Fromm, and L. Kohlberg—describe the mature person as freed from unquestioning obedience to society's conventions and rules. The mark of healthy people in psychological thought is the ability to reason and to create their own decisions about morality and behavior. This view is addressed in many different conceptions of psychological health such as White's view of individual competence and Rotter's view of internal locus of control. At the lower end of the scale of psychological development, most psychologists place individuals who are unable to think or act for themselves because of learned helplessness, low self-esteem, and neurotic dependencies. Basically, then, poor mental health is seen to arise from an inability to control one's own behavior.

Autonomy, then, is regarded as an indication of maturity and psychological health in Western society. The goal of almost all psychotherapies is to restore behavioral freedom and remove stifling mental controls. As I have mentioned, psychologists Hare-Mustin and Marecek[10] express reservations about advocating autonomy as a goal of therapy, particularly for women. Nevertheless, they point out that there are several ways in which individual autonomy contributes to mental well-being. First of all, autonomy implies individualism and self-sufficiency. Autonomy involves the sense that one has separate and legitimate needs which one is justified in pursuing. Second, autonomy implies power to make choices, to determine one's best interests, and to take actions and make decisions. Influence as well as mastery and competence are important aspects of autonomy. Finally, autonomy allows one to determine independence from others around oneself.

From many modern perspectives then, the autonomous individual is valued. The perception that individuals are free to choose their own outcomes has traditionally been associated with positive appraisals of their worth. To the contrary, the person who cannot control his or her own behavior is seen as less worthy. At best, we pity and wish to help the crippled, the homeless, the mentally ill, and all such persons who seemingly have lost control of their lives. At worst, we view them with contempt, blame them for their lack of control, and in all ways avoid social contact with them.

The Consequences of Using Strong Tactics

The successful use of strong influence tactics creates the belief that people are not in charge of their own behavior. I described in Chapter 2 how the control of resources needed by others provided the individual with power, and encouraged the use of strong tactics of influence. Chapter 2 also described the reciprocal relation between attributions of control and the use of strong tactics in marriage relations. Spouses who "demanded" and "insisted" also believed that they had the final say when disagreements arose. However, since these findings are based on survey data, they provide no direct proof that the use of strong tactics can change beliefs about the autonomy of the person being influenced. It is just as likely that strong tactics were used because individuals had power to control their spouses.

To examine whether the use of strong tactics can change subsequent beliefs about the target of such influence, my students and I[12] conducted a series of experiments. In these experiments we created a business and appointed students as managers. These student managers were delegated the use of influence tactics of varying strength in order to supervise their employees. Through this experimental procedure we could measure changes in beliefs about employees' autonomy in relation to the strength of the influence tactics that the managers used.

To illustrate, in one of these industrial simulation studies, one-half of the student managers were allowed to use a broad range of strong tactics when supervising their employees. That is, they could give pay raises, deduct money from their employees' salaries, fire them, shift their work, and send them for extra training. Thus in all ways these managers had strong tactics of influence available to control their employees. A second group of student managers were simply told that they were the bosses. However, they were given no delegated powers to help them influence their employees. Thus, if they wanted to influence, they were limited to simple requests and discussions.

One of the first findings of this study was that managers who were allowed to use strong tactics were far more active in giving orders than managers who could only discuss and request. That is, managers with delegated powers were observed to issue almost twice as many orders as managers who did not control such powers. As I pointed out in Chapter 2, the control of resources that others give weight encourages the exercise of influence. Influencing agents want more from others when they believe they can get more (i.e., force others to comply).

Not being dependent on the good will of followers for compliance – not having to ask and discuss and wait for the decision of followers – appeared to encourage managers in this instance to increase the number of demands made of their employees.

We next examined whether the use of strong tactics changed the managers beliefs' about their workers' autonomy. At the completion of the actual work, the managers were asked to evaluate their employees' performances. One set of questions asked the managers to estimate what had caused their employees to perform effectively. There were three alternatives, and each was evaluated as to its importance: "The workers' own motivations to do well"; "My orders and suggestions"; and "The workers' desire for money."

Managers who could only discuss changes with their employees attributed their employees' effective performance to the employees' own motivations to do well. Managers who used strong tactics (i.e., rewards, threats, changing the employees' jobs, training them, etc.) stated that the employees' effective performance was due to their orders and suggestions. The use of strong tactics, then, led to the belief that the employees were not performing autonomously. Rather, their behavior was regulated by the powers controlled by the manager.

And so it goes at all levels of behavior. From the young child bullying other children or begging for compliance, to statesmen issuing ultimatums or negotiating settlements, the strength of the influence tactics that are used alters the extent to which those being influenced are viewed as in charge of their own behavior. Strong tactics are seen as forcing compliance. Weaker tactics are seen as allowing other people the freedom to decide for themselves. Stating this relation in terms of attribution theory, as this theory is stated in psychology, the use of strong tactics causes the perception that the target person is externally rather than internally controlled.

The autonomous individual is evaluated more favorably than individuals who are seen as not in charge of their own behavior. We move away, socially and psychologically, from individuals who are seen as not in charge of their own behaviors. The thesis of the metamorphic model of power is that when people are seen as not responsible for their own behaviors, influencing agents see them as less worthy than themselves, express the desire to move away from them socially and psychologically, and in many instances carry out acts of harm against them that would never be considered when dealing with persons of equal power. One reason for this, as I have suggested several times, is that credit for acceptable behaviors is only given to persons who are seen as in charge of their own behaviors.

Control and Affection

We turn now to empirical evidence to support the above generalizations. One set of such evidence is found in studies of the relation between autonomy and evaluation among men and women who are married or engaged. We have found invariably that when one person believes that he or she makes all the decisions in the relationship (i.e., controls the partner's behavior), that person also evaluates the submissive partner unfavorably.

To illustrate, my colleagues and I (see Kipnis, Castell, Gergen & Mauch[13]) asked 76 married men and women about the degree to which they shared power with their spouses in terms of having the "final say." We also asked the spouses to evaluate their partners in terms of various traits and abilities such as: capability of solving problems; skill in the work they do; common sense, reliability, intelligence, and other evaluative traits. To measure emotional and psychological closeness to their spouses, we asked the respondents to tell us how happy they were in their marriages. Our measure consisted of statements about their marriage to which respondents agreed or disagreed such as: "I have never been happier" and "Our marriage has not worked very well." A second measure of psychological distance was in terms of the respondents' satisfaction with their sexual relations with their spouses based on the question: "Our sex life is: outstanding; excellent; fine; average; somewhat below average; poor."

The findings were that our measure of dominance correlated $-.42$ with each person's evaluation of his or her spouse. The more respondents reported that they controlled decision making in the family, the less favorably they evaluated their submissive spouses. Thus, husbands, or wives who continually decide what their spouses should do are forced to conclude, ever so unwillingly perhaps, that their spouses do not measure up as capable, skilled, and intelligent people. Why should they? All they are doing is what their spouses told them to do.

Perception that the spouse lacks autonomy was also related to an expressed desire to increase social distance and move away from the submissive spouse. In our survey, we found that the dominant spouses expressed less satisfaction with their marriages and less enjoyment with their sexual relations with their spouses. That is, dominant spouses were more likely than those who shared decision-making powers equally with their partners to endorse such statements as "I have found little happiness in our marriage," and "our sex life is poor or below average."

Similar findings on dominance, evaluations, and affections for one's romantic partner have been found in several other studies among married and dating couples carried out by me and my students. In all instances, when power was unbalanced in the relationship (i.e., when one person made all the decisions for both), the more powerful person held little affection for the submissive partner. Taken together, these findings lead to the observation that love and dominance are incompatible partners. Macho individuals, for instance, whose goals are to dominate and control their partners, will continually find disappointments in love relations, but not necessarily because their submissive partners resent being controlled. Indeed one frequently finds that such control represents security, rather than repression, to the submissive partner. Rather, love fades because the machos' attention wanders to new challenges as they find little to hold their interest in their submissive partner behavior. "All she does," goes the complaint, "is what I tell her to do."

Control and Evaluations at Work

We have also found much the same pattern of relations between control and evaluations of the target person in our simulations of organizations. It can be recalled that, in these studies, some managers were given a broad range of powers (rewards, coercion, ecological control, and training) to influence their employees, and other managers were given no power beyond telling them they were the boss.

At the completion of the actual work, as I described in the previous section, it was found that managers who used strong tactics (i.e., rewards, threats, changing the employees' job, training them, etc.) believed that their employees' effective performance was due to their (the managers') orders and suggestions. The use of strong tactics, then, led to the belief that the employees were not performing autonomously.

We also asked the managers to evaluate their employees' job performances. It should be noted that the employees were confederates of the experimenters. Hence, they produced the same amount of work in both experimental conditions. Despite the fact that their performances were identical, workers whose performance was attributed to the manager's orders received far poorer evaluations than workers whose performance was attributed to their own motivation to do well. That is, autonomous workers were described by their managers as more competent, as more deserving of promotion, and more deserving of rehire than workers whose performance was attributed to the managers' orders. Thus the same employee performances were evaluated differently depending on the strength of the tactics that the managers used and the managers' subsequent attributions about worker autonomy. My explanation is that managers with power simply assumed that their orders had caused their employees to work effectively. Hence they could discount the worth of their employees' own contributions to the work.

The ease with which power transforms attitudes and beliefs was also evident from the managers' comments at the end of the study. The managers' were asked what they had to do in order to be successful in their roles as managers.

Managers without power continually stressed the importance of allowing workers freedom to do work on their own. By being allowed this freedom, they said, the worker would be motivated to perform at high levels. For example, one manager without power said, "You must have control, but not to the point where you would dominate the worker. You must also have gentleness so you wouldn't offend the worker." Another manager without power said, "You should have ability to show confidence in the worker, encouragement, and allow them freedom to perform their jobs in their own way."

The use of strong tactics, however, transformed the attitudes toward their employees of the remaining managers. Those who had been given power stressed the necessity for added control in order to raise production: "You have to know how to influence the men to do more and to do it better." Another manager said, "You have to know how to motivate the workers, even when they may not want to be motivated." An analysis of these statements found that 76% of the managers with power and 21% of those without power expressed manipulatory attitudes about their workers.

Thus, in this instance, the control and use of strong tactics created what may be termed an authoritarian climate in which managers came to view their employees as objects of control, unlikely to work unless forced to do so by the manager's orders. Further reinforcing this climate was the attempts by managers who used strong tactics to "move away" socially from contacts with their employees. When asked at the end of the simulation, managers with power were less willing "to have a cup of coffee or Coke with their employees" than managers without power. These findings reflect, I believe, our indifference to people who are seen as not in charge of their own behavior. To be controlled by another person robs one of those very qualities of dignity, competence, and self-worth that attract one person to another.

Does Power Corrupt?

The answer I have suggested in this chapter to this question is that it depends on the tactics that we use, and the attributions about control that we make. When compliance-gaining strategies are used that allow those being influenced to decide for themselves, social relations remain cordial. To the extent, however, that getting our way means depriving others of behavioral freedom, then psychological changes are produced in the influencing agent of the kind described in this chapter. Under such circumstances, we exchange contempt rather than gratitude for compliance. As a result, social relations between those who are controlled and those doing the controlling change for the worse.

In the following chapters I shall examine these ideas as they apply to technology. The theme is much the same as described here. To the extent that technology increases the ability of actors to control outcomes that involve people, or to the extent technology takes control away from people, we can expect social relations to follow the course described above.

References

1. Zimbardo, P.G. (1969). Individuation, reason and order versus deindividuation, impulse and chaos. In W.J. Arnold and D. Levine (Eds.), *Nebraska symposium on motivation* (pp. 237–307). Lincoln: University of Nebraska Press.
2. Wiser, W., & Wiser, C. (1967). *Behind mud walls*. Berkeley, CA: University of California Press.
3. Rogow, A.A., & Lasswell, H.D. (1963). *Power, corruption, and rectitude*. Englewood Cliffs, NJ: Prentice-Hall.
4. Scott, J.G. (1972). *Comparative political corruption*. Englewood Cliffs, NJ: Prentice-Hall.
5. Sorokin, P.A., & Lundin, W.A. (1959). *Power and morality: Who shall guard the guardians?* Boston: Sargent.
6. Szulc, T. (1986). *Fidel: A critical portrait*. New York: Morrow.

 7. Sampson, R.V. (1965). *Equality and power.* London: Heineman.
 8. Smith, T. (1987). *Thinking like a communist.* New York: W.W. Norton.
 9. Thibaut, J., & Walker, L. (1975). *Procedural justice: A psychological analysis.* Hillsdale, NJ: Erlbaum.
10. Hare-Mustin, R.T., & Marecek, J. (1986). Autonomy and gender: Some questions for therapists. *Psychotherapy, 23,* 205–212.
11. Becker, E. (1968). *The structure of evil.* New York: The Free Press.
12. Kipnis, D. (1981). *The Powerholders* (2nd ed.). Chicago: University of Chicago Press.
13. Kipnis, D., Castell, P., Gergen, M., & Mauch, D. (1976). Metamorphic effects of power. *Journal of Applied Psychology, 61,* 127–135.

Part III   Technology and Control

We can now proceed to the principal concern of this book. In the following chapters we look at how technology, in such diverse areas as medicine, work, the military, and the social sciences, changes the values and attitudes of persons who assume control of these technologies. Each of these controllers has been given, in large measure, power to determine outcomes that involve other people. This added power has come, not from the magic wands of the gods, but from the application of pure reason.

Part III. Technology and Control

CHAPTER 4

Behavior Control Technology

The "Relax and Learn System" uses the psychological technique of
Multisensory Emotional Therapy *in order to change a patient's thinking,*
attitudes, and behaviors. You can begin treating your patients immediately
using this innovative and proven effective technique.... This system
includes an IBM compatible computer with the appropriate video boards,
one 20 mg hard disk with "Relax and Learn" software already installed, one
floppy disk drive, and an excellent special monitor with speakers. This
approach is not only cost effective for the patient, but ... is also cost
effective for the psychologist as well. Since this technique does not require
him to spend the entire therapeutic hour with these patients, it is an
excellent way to expand his practice without increasing his overhead
significantly, as well as increasing his cost effectiveness.

Write to: Advanced Psychological Technology—For Stress, Pain, Cognitive
Control and Attitude Change Profiles International, Inc. Fort Worth, Texas.

Since Freud's initial discoveries about the treatment of anxiety, techniques to
control behavior have been developed in an astonishing number of areas of human
activity. As a result, practitioners of psychology are experiencing a golden age.
Their services and advice are sought and accepted in practically all fields of
human activity. Newspapers describe the activities and opinions of psychologists
on marriage, love, child rearing, and other aspects of day-to-day life. In the fields
of marketing, personnel, training, selection, and more, executives rely on the
advice and opinions of consulting psychologists.

What is new about psychological knowledge and gives it popular acceptance
is that its findings are based on scientific discoveries, rather than, say, divine
revelation or the peculiar fancies of recluse professors writing about how people
should behave. People know that psychology uses the scientific method to evaluate the correctness of its ideas and methods. As a result, few psychologists
are seen as quacks selling what they cannot produce, or as Svengalis or Ras-
putins using animal magnetism to manipulate unwilling victims. Rather, psychologists are presumed to have a deep understanding of human nature, a range
of proven technologies, a profound concern for people, and a seeming disinterest
in their own aggrandizement. The general belief is that psychologists are here to
serve and help.

Behavioral technology, then, has empowered psychologists. What people want
to know, and what psychologists appear to know, is how to change behavior.
People ask psychologists to help them lose weight, stop smoking, understand
themselves, be more persuasive, or stop feeling anxious, angry, or depressed. Or
perhaps they want to learn how to make friends, be better salespersons, change

other people's attitudes, communicate persuasively, win elections, make others obey, or work harder. Businesses want to know who to hire, who to fire, and who to promote. Firms pay psychologists for advice concerning how to reduce conflict, to improve employee morale and motivation, and to figure out what people like or dislike about their products. Parents ask psychologists how to socialize their children to be achieving, affectionate, and happy members of society. College students want to know which careers will allow them to make a real impact on the world, and on other people. In short, people want psychologists to tell them how to be successes; or if not that, how to avoid failure.

This is a tall order. In earlier times, people would have struck a bargain with the devil for even half of these outcomes. But psychologists say they can provide the answers, and even more. And, seemingly, without people losing their souls.

The reason for psychologists' confidence is that their methods, as I mentioned, are wrapped in the glow of science. Over the last 50 years, psychology has developed and validated a variety of behavioral technologies that can be used to alter behavior, with modest degrees of certainty. By modest I mean that the application of a given technology changes a greater proportion of people's behavior than could be expected by chance. Of course, not everyone changes. But enough people do change to convince skeptics that the payoffs of using psychological knowledge outweigh the costs involved. Proven behavioral technologies, then, are the basis for psychologists' newly found power to influence.

What Is Behavioral Technology?

Perry London[1] has described behavior control as one's ability to get someone to do one's bidding. The techniques of psychotherapy, for instance, widely practiced and accepted as a means of curing psychological disorders, are also methods of controlling people. Behavioral technology, in most instances, consists of instructions about how to present verbal and nonverbal stimuli to persons we wish to influence. Usually these instructions are contingent upon information about the setting and/or the characteristics of the recipients of influence.

In other instances, behavioral technology provides instructions about how to alter the environment of the recipient, and so change his or her behavior. In still other instances, behavior technology combines the use of verbal and nonverbal stimuli with environmental changes. Finally, there are surgical and pharmacological agents that are used to alter behaviors. Most frequently, the latter change agents are used to control mental illness, such as depression and schizophrenia. However, Henry Clark[2] has recently argued for a form of "technological humanism," based on the artificial stimulation of the nervous system and the use of chemicals to increase sensory experiences and alter mood states. Such techniques, he argues, will promote happiness—surely the ultimate technological "fix."

Regardless of the methods employed, the goals of behavior technology are the same as those of all technologies. That is, to produce behaviors that are predictable and controllable from the perspective of an influencing agent, for example, advertiser, supervisor, psychotherapist, parent—people who have to influence others to achieve practical outcomes. And, as is true of most technologies, attempts to reduce behavioral uncertainty often involve reducing the amount of control and free choice available to the target person.

Of course not all behavior technologies are designed to reduce free choice. Thus, training supervisors to use participatory leadership, or conducting psychotherapy with Rogerian or existential therapy, are forms of behavioral technology that are designed to increase the target person's free choice and autonomy. Yet, when one examines the full range of behavior technologies that have been developed, it is fair to say that most are designed to reduce individual choice in order to produce the desired behaviors.

Because of the requirement that behaviors be controlled, behavior technology and power are closely related in several ways. First, power is involved because technology restricts the range of individual behaviors that can be displayed. By this I mean that behavior techniques guide people so that they behave in some ways, but not in others, or think and feel in some ways, but not in others. Second, power is involved in that behavior technologies are developed that are consistent with the dominant political and economic interests of society. That is, techniques of behavior control are supported and find acceptance to the extent that they help bind the individual to existing institutions, such as the family, business, government, and the military. Techniques that successfully cause opposition to these institutions are unlikely to gain wide support. Techniques, for example, that encourage teenage rebellion or drinking would find few buyers.

Beginning with World War I, government and industry have provided major support for the development of behavioral technologies in such areas as psychological tests, attitude measurement, leadership training, attitude change techniques, and learning. In the area of public health, government and private agencies have supported the development of techniques to reduce delinquency or substance abuse, and to increase the effectiveness of various forms of psychotherapy. In short, the techniques developed by psychologists serve, not unreasonably I believe, to strengthen rather than to weaken existing social institutions and normative values.

Behavioral Control Technologies

As I mentioned, technologies that guide human behavior have been developed in an astonishing number of areas of psychology. Some techniques attempt to control very specific behaviors, such as littering. Others have more general application and attempt, say, to improve children's abilities to learn, or to reduce conflict between distressed couples. In this section I describe some of the major technologies in current use. Not all technologies are described. Such a compendium would be too long. Rather, the purpose of this listing is to illustrate the various ways in which behavioral technologies transfer control away from target persons, and so increase the influencing agent's ability to predict and control behavior.

Psychological Testing

Perhaps the oldest and most successful behavioral technology available today is the use of psychological tests to select persons for training and for work. Tests control uncertainty by denying access to persons liable to do poorly in training or at work. The mean level of performance of the selected applicants is raised by eliminating poor risks, while at the same time the variability of such performance around this new mean is reduced.

Another benefit for users of tests is that time and effort do not have to be devoted to influencing the performance of these "poor risks." Instructional and managerial time can instead be directed toward improving the already acceptable performance of those selected by the tests. Thus, in addition to the economic benefits of denying access to less talented persons, there are social psychological benefits in terms of conflict reduction.

Ecological Forms of Control

Rather than attempting directly to alter a person's thinking, feelings, or behavior, one can control behavior by changing the person's physical or social environment. This form of control is based on the Darwinian principle that changes in the environment encourage changes in behavior. Psychologist Dorwin Cartwright[3] called these environmental interventions "the exercise of influence through ecological control." Technology based on ecological control is remarkably unobtrusive because people often adapt to their surroundings without awareness of the environmental processes that have guided changes in their behavior.

Ecological control techniques cover a broad range of interventions. At the micro level, there are programmed learning devices that teach new skills by systematically controlling the range of responses their users can make. At a macro level, there are large-scale alterations in the individual's environment such as are found in scientific management techniques (e.g., work simplification), or in psycho-architectural techniques for altering interior design or room shapes to facilitate or restrict conversations between people.

As the reader can appreciate, face-to-face confrontations between target persons and influencing agents can be avoided when ecological control methods are used. New automated factories can be built in which the need for employee skills is reduced, and furniture can be arranged before patients enter a room. Once they are established, these techniques allow influencing agents to operate behind the scenes. The target person's behavior is seemingly guided by the environment rather than by the direct influence of another.

Nonverbal Influence Techniques

Individuals produce and respond to a variety of nonverbal signs that signal approach, avoidance, readiness to converse, greetings, flirtation, aggression, dominance, submission, turn taking in conversations, and many other behaviors too numerous to mention. These influence techniques are conveyed through the emotional aspects of speech, the use of personal space, and expressive movements of the body. Aspects of personal grooming, displays of prestigious possessions, and interior decorating also serve as nonverbal displays capable of influencing behavior.

How can knowledge about nonverbal behavior be used to control others? One way is to teach people how to use nonverbal displays. Teachings can range from parental admonitions such as "Don't slouch" and "Look people in the eye," to the development of commercial programs such as neurolinguistic programming. These commercial programs teach people how to use their own body language to influence others silently. They also provide instruction about how to decode the target person's body language and use this information to influence. As in the use of ecological control methods, this behavioral technology can be used without the target person's awareness.

Behavior Modification

This technology is limited mostly to hierarchical interactions in which more powerful persons attempt to modify the behavior of the less powerful through control of sanctions, mainly rewards, that are meaningful to the individual. Behavior modification is used either with the consent of the target person, as for example in helping people control their eating habits, or without their consent, as for example in allowing supervisors to control employee performance. In all instances, the technology is based on careful planning in which the powerholder identifies the behavior that is to be controlled, the environmental circumstances that promote or inhibit the behavior, and the contingencies between these sets of events.

Homeostatic Mechanisms

Psychological states of imbalance, or dissonance, has been found by psychologists to change an individual's beliefs or behavior in directions advocated by the change agent. By inducing the discomfort of dissonance, a wide variety of behavioral and cognitive changes have been effected, ranging from causing people to enjoy boring experiences to long-term changes in belief systems. The use of this technology to control behavior has been advocated in such diverse areas as sales, psychotherapy, and marketing.

Verbal Influence

The words that people use to persuade are the most common means of exercising influence. People beg, plead, challenge, demand, ask, and discuss far more often than they use selection tests, ecological control, behavior modification, body language, or dissonance. Books about verbal tactics that work (e.g., Cialdini,[4]) are read by politicians anxious to convince their constituents, by young executives who want to be promoted, by lonely bachelors looking for a mate, by advertising agencies who want to attract a larger segment of the market, and by harried parents trying to learn the best way to control their children, to name but a few of the many people who want to learn the right words.

It is not surprising, then, that behavioral technologies based on persuasive influence tactics are found in all the applied fields, including leadership training, jury selection, marketing, advertising, and psychotherapy. Their use substitutes science for intuition. Books about how to get your way are psychology's update of yesterday's juju beads, hex signs, and voodoo spells.

Most modern systems of verbal influence have a common core of procedures, although the names given to these procedures vary. In industry, for instance, the procedures may be called contingency leadership, and in psychotherapy cognitive restructuring. In almost all of these instances the influencing agent is provided with the means to diagnose the situation and/or the psychological state of the person being influenced. Using this information, the influencing agent can then invoke the particular combination of words that have been found to be effective in that situation. For supervisors, these right words may consist of orders whose content varies in terms of task or socio-emotional orientation, whereas in a public health media campaign, the message might vary in terms of anxiety provoking content or one- or two-sided messages.

As can be seen from this brief survey, psychological research provides influencing agents with a wide range of behavioral technologies. Some are quite effective in reducing behavioral uncertainty. This is particularly true of technologies that alter environments, such as task routinization or changing architectural space. Target persons find it particularly difficult to reshape fixed features of their environment. Other technologies appear to have minimal long-range effects and control behavior only when actively applied, as for example, influence strategies based on the use of sanctions.

A major goal of applied psychological research is to provide information about the effectiveness of the above technologies in reducing behavioral uncertainty. The literature contains considerable information about the extent to which these technologies work. However, almost nothing is known about the impact of the use of these technologies on the influencing agent. In the remaining sections of this chapter, the findings of empirical studies will be presented that suggest that the use of behavioral technologies changes influencing agents in predictable ways. But first I will examine ethical concerns that may be voiced about the propriety of controlling behavior.

The Ethics of Behavior Control

When I get too concerned about ethics, I think about what would have happened if Alice had been warned about the consequences of drinking the bottle labeled "Drink me." Of course, ethical concerns would have been satisfied. Alice would have been given a choice, and she would have known what a "mess" she was going to get into. Perhaps she would have drunk anyway. What worries me is if she had decided not to drink. Lewis Caroll would not have had much of a story.

Given the close relation of behavioral technology to issues of control, there are a number of ethical concerns that can be raised about its use. Normally such concerns may seem unreasonable to psychologists, who, if they think about it at all, believe that behavior technologies are developed to help people rather than harm them. While codes of ethics may be needed to protect people from harm, they are not needed to protect people from being helped. Furthermore, in many areas the person who is to be influenced must give his or her consent before control techniques can be applied. Thus, for instance, people freely apply to have such therapeutic techniques as "desensitization," "flooding," and "deconditioning" used to help them, for example, eliminate fears, stop smoking, or face life more squarely.

On the other hand, many behavior technologies are developed and used that do not require the consent of those who are to be influenced. For example, at the 1988 meetings of the International Congress of Psychology, papers were presented with titles such as: "When and how message effectiveness can be increased by matching appeals to recipients' personalities" and "The effects of different communications and type of authority on the motivation to volunteer for officer training." Presumably the application of the findings of these papers – to make people believe persuasive communications, or to increase the rate of volunteering – would not involve asking recipients for permission first, or obtaining their consent to be the subject of such influence attempts. Permission would not have been sought because the behavior technologies were developed by parties whose goals differed from those who were to be influenced. It is hard to imagine an advertising firm asking consumers for permission to influence them.

Thus, ethical issues are clearly involved in the development and application of behavioral technologies. Several writers have expressed concerns about the possibility of using psychology's findings to exercise political control. Social scientists, such as Freedman and Freedman,[5] as well as psychiatrists, such as Seymour Halleck[6], have described the ethical issues that result from the use of applied psychology to restrict and guide individual political behavior.

In his book *Behavior Control*,[1] Perry London has broadened the scope of this discussion by suggesting that the deliberate control of human behavior is immoral simply because it dehumanizes human beings. London bases his argument on the assumption that individual autonomy, as discussed in Chapter 3, has absolute priority in making decisions about human behavior. Anything that reduces an individual's ability to make choices (whether he or she wants to make choices or not) is objectionable precisely because it does so. London argues that the exercise of choice is the heart of morality, which in turn is one essence of humanity. Since the imposition of control is the very antithesis of choice, it, ipso facto, dehumanizes. And since humans are morally obligated, above all else, to exercise their human attributes, they should not support an enterprise dedicated to the subversion of such attributes. In short, London offers the ethical rule that people should neither be forced or seduced into compliance. Rather, people should be free to decide for themselves how they want to guide their lives.

A second concern of London's worth repeating here is that we share with animals behavior that can be manipulated. But London argues that animals do not have the moral imagination to consider such matters as good and evil—only humans do. To treat an animal as though it possessed this capacity is merely silly, but to treat human beings as though they were only their behavior is wicked.

The issues of free choice and the manipulation of behavior, then, are seen to underlie the ethical dilemma of behavior technologies. The very possibility of deliberate behavior control is frightening because it conjures the Orwellian vision of humanity's future. Yet it is also true that the principle of individual free choice as the ethic guiding human experimentation, when taken to its extreme, serves to paralyze the application of social science findings. From this extreme position, one ends by defending the view that all technologies that restrict choice are unethical. Because technology does limit free choice, to a greater or lesser degree, we are reduced to the position that technology is unethical. From this perspective, only "man in nature," unfettered by the restraints of civilization, is free. This extreme position is obviously silly in today's society. The social and ethical problems generated by technology will not be solved by eliminating all technological knowledge.

However, in its less extreme form, this concern for human choice alerts us, as is London's intent, to the many ethical dilemmas involved in the use of behavioral technology. At a minimum, London points out that behavior technologies lull our ethical sensitivities by proving that there are many exceptions to absolute ethical laws. Thus, for example, the precept "Thou shalt not kill" is watered down by social science findings that state that many people cannot be held responsible for their own acts.

From my perspective, current discussions of the ethics of behavior control are incomplete in that they say little about the effects of technology on those who use it. Yet changes in the influencing agent, as a result of controlling others' behavior, have ethical implications. These changes, as I will show, involve the loss of a moral perspective in dealing with people, transforming one person's perception of another from person to object. Such changes are, I believe, as immoral as restricting free choice. If the individual who is to be controlled has a right to be advised of the risks involved, so, too, should controllers have similar rights. An evenhanded set of ethics suggests that awareness of the side effects of using technology should be made available to all parties involved.

Can the Metamorphic Model Be Applied to Behavioral Technologies?

For the most part, psychologists are humanists, and the techniques they develop are designed to help people, not to hurt them. Thus, it may come as a surprise to find that many of these techniques may have the unexpected consequence of causing users of behavioral technologies to derogate the very people they are trying to help. This outcome may occur because, if we are to change behavior, the techniques we apply must in some way control behavior.

Unfortunately, psychology has not been particularly concerned with examining possible metamorphic effects of behavioral technologies. The psychologist, as controller of behavior, remains a benign enigma in all eyes, including psychologists. Questions about the misuse of psychological knowledge focus on those who are exploited or on the relative frequency with which exploitive behaviors occur.

Misuse of Psychotherapy

In a nationwide survey of U.S. psychiatrists, Gartrell and her associates[7] reported that 7.1% of the male and 3.1% of the female respondents acknowledged sexual contact with their own patients. Similar results have been reported in surveys among clinical psychologists. Such contact is explicitly prohibited by both psychiatrists' and psychologists' codes of ethics because of its harmful effects on patients. Nevertheless, a small proportion of therapists report that they do become sexually involved with their patients. That this number may be higher is suggested by the Gartrell et al. finding that 65% of all psychiatrists surveyed reported treating patients who had been sexually involved with previous therapists. Either the roughly 5% of psychiatrists who admit sexual relations with their patients treat a very large number of patients, or other unethical psychiatrists chose not to answer Gartrell's questionnaire.

In his book *Beyond Words*, Kurt Back[8] has described how the encounter movement and sensitivity training also places participants at risk as a result of the group leader's behavior. This movement uses a behavior technology— initially developed by Kurt Lewin and his associates, and further refined by the National Training Laboratory, Bethel, Maine, and others—in which feedback from small groups is used to control the behavior of individuals within the group. The goal of training is to encourage individuals to discard their conventional defenses and ways of coping. It is assumed that personal learning and psychological growth will result from the insights gained during the group experience.

There are relatively few codified ethical controls placed on encounter group leaders in comparison to those placed on psychiatrists and clinical psychologists. There are several reasons for this absence of codified ethical controls. First, sensitivity training is not necessarily considered psychotherapy. Sensitivity training can be found in management training sessions, in sessions offering personal growth, and even in singles groups as a means of getting people acquainted. Thus, strong formalized association codes of ethics are frequently considered irrelevant. Second, the techniques of the encounter movement are relatively easy to learn. As a result, many lay persons can become practitioners without belonging to formal organizations that might regulate their practice. Third, an assumption of the encounter movement is that each participant is responsible for taking care of him- or herself. Thus, the role of the leader is in theory to be passive, letting each member do his or her "own thing."

In practice, of course, encounter leaders have considerable power in the group. Their actions or inaction serve to cue the group as to what behaviors toward each other are acceptable. Furthermore, it is the leader who initiates group activities. Thus, the leader can instruct participants to challenge each other's beliefs and life goals, to become aggressive, to reveal intimate information about the self, to touch each other, and to use group exercises that provide sometimes unwanted group feedback about the self. The many techniques developed by the encounter movement provide leaders with considerable power to control the behavior of group members.

The problem is that the strong emotional experiences resulting from group feedback cause psychological casualties. For example, a controlled study by Lieberman, Yalom, and Miles[9] showed a casualty rate of over 9% due to sensitivity training. What may happen is that the most angry and aggressive members of a sensitivity group can use the group exercises to attack weaker group members, disguising their hostility as objective feedback.

Back[8] attributes these psychological casualties to the failure of the group leader. Some leaders may allow vulnerable group members to be psychologically attacked because of the mistaken idea that they are not responsible for what happens. The reality, he points out, is that group leaders are perceived by the group as being in charge, and their failure to halt destructive group activities is seen as a signal to continue. Other group leaders are seen as simply abusing the power they control. Thus, in discussing the harm done by sensitivity training among management groups, Robert Kaplan[10] argues that

the power advantage allows leaders to show their sadistic side and get away with it. Leaders can cloak their destructive behavior in a mantle of righteousness. They act as if their training and their role require them to be demanding and harsh. (p. 765)

Using Psychological Knowledge – Effects on the User

The above cited findings, when taken together, suggest that the use of psychological knowledge has a destructive side. The findings tell us that certain numbers of therapists violate the ethical standards of their associations, or that the use of behavior techniques can cause psychological casualties. What is missing however are systematic attempts to account for these ethical failures. One wonders whether such failures represent only the tip of the iceberg. The metamorphic model, for instance, suggests that there should be changes in the social attitudes of most social science practitioners, not necessarily in the unethical ways reported above, but in terms of how they regard those they control.

By this I mean that with behavior technology, psychologists no longer have to beg, act polite, or simply ask. They can cause behavior by using the right combination of words, by using reinforcements in strategic ways, or even by changing others' environments. Although these increments in social power may make those who are influenced more compliant, the metamorphic model predicts that added control will make influencing agents, at least more aloof from others. We move rapidly from a posture of gratitude to one of easy indifference and worse as we become the origins of other people's behavior.

Unfortunately, there are only a few reports of changes among social science practitioners, and these are limited to self-reports by psychotherapists about changes in their personalities that appear to result from psychotherapeutic practices. We know nothing about similar changes among other applied practitioners in, say, industry, politics, or advertising.

In a recent review of these self-descriptions, James D. Guy and Gary Liaboe[11] reported that therapists describe themselves as becoming more assertive, self-assured, self-reliant, introspective, and sensitive. They also described themselves as feeling aloof and distant from their family and friends. As a result, they reported having a smaller circle of friends and socializing less during a career of conducting psychotherapy.

Guy and Liaboe suggest that many of these changes may stem from therapists' ability to change other people's behavior. Because of this ability, therapists receive admiration and respect from patients and the community. Furthermore, many people are in awe of therapists' almost extrasensory powers – their ability to see through people's defenses and solve their psychological problems. This idealization may cause the therapist to feel superior and to consider him or herself an "expert." As a result of these feelings of superiority, therapists' own self-esteem is raised and at the same time they "move away" from lesser individuals. While this explanation is consistent with the metamorphic model, Guy and Liaboe also point out that because these changes do not happen to all psychotherapists, there are many alternate explanations for the reported changes, including personality differences among therapists, differences in the therapeutic training they received, and/or the therapeutic techniques they use.

In addition to the above studies among psychotherapists, I am aware of only one other published report that examines how the use of behavior technology could change those who used it. In an experimental study, Dutton[12] found that the use of attitude change technology altered subsequent attributions of the influencing agents regarding the extent to which they controlled the attitudes of other people. Dutton found that influencing agents who were told they were using scientifically validated persuasion techniques (i.e., two-sided arguments, support of high status persons) believed that their arguments had forced listeners to change their beliefs. On the contrary, agents who were told they were not using scientifically validated techniques attributed subsequent changes in beliefs to the target person's own decisions to change. That is, the target person thought over the information and decided that the agent's ideas were correct. In this instance, then, the use of attitude change technology, rather than the free choice of the target person, was seen as the cause of the target person's new beliefs.

In sum, the available literature suggests that the control of behavior technology, particularly therapeutic techniques, may be associated with changes in those who use the technology. However, the amount of information about such changes is very limited. In addition, with the exception of Dutton's experimental study, the information is ambiguous and subject to alternate explanations. We do not know whether the very use of behavior technology caused some therapists to exploit their clients and other therapists to move away from friends, or whether these changes were the result of pre-existing personality differences among therapists.

These brief descriptions highlight the question of whether the use of behavior technology changes the influencing agent. To provide some information about this topic I have carried out several empirical studies. These examined how the use of leadership technology affects leaders' evaluations of their followers and how the use of psychotherapeutic techniques affects similar evaluations of patients by therapists. In both of these studies, I contrasted the use of behavior technology designed to restrict the target person's autonomy with behavior technology designed to increase such autonomy.

Therapeutic Practices and Changes in the Therapist

An assumption of many mental health practitioners is that patients are not aware of the causes of the oppressive stress in their lives. Thus, the task of psychoanalytically oriented therapists is to provide patients with insight about the sources of the stress they are experiencing. This is done through such techniques as reflective listening, encouraging the patient to connect what is happening within the therapy session to his or her life stress, occasionally providing explanations about the patient's behavior that differ from the patient's own explanations, and sometimes even frustrating the patient so that the patient learns about how he or she reacts to frustration and hostility. It is generally agreed that dynamic therapy is a long and uncertain process whose success depends upon the ability of the patient to understand and eventually change his or her behavior.

Seymour Halleck,[6] a professor of psychiatry at the University of Wisconsin Medical school, describes an illustrative case of psychoanalytic therapy:

A thirty-year old professor became severely depressed when he began to experience work blockages. By the time I saw him he was spending most of his days at home, unable to work. He was talking of suicide.

The patient was seen in psychoanalytically oriented psychotherapy three times weekly for a period of three years. The first year was extremely stormy ... eventually however the patient gained insight into his difficulty. He used the therapeutic relation to learn that hostility could be expressed with some safety and that affection could be obtained if he behaved in a mature and confident manner. These therapeutic gains gradually affected his life situation. He learned to resist his employer's exploitation of him. As he became more assertive at home, his wife appreciated his determination and eventually was able to provide much of the love and nurture he needed.

In this case, then, [Halleck concluded] *the patient had the strength to use his insight into past and present conflicts to enable him to deal with his environment.* (italics added) (pp. 57–58)

The assumptions of more recently developed humanistic schools of psychotherapy also stress the importance of patients' discovering for themselves the basis for their discontent. However, humanistic schools favor spontaneous expressions of feelings, flexibility, and personal relationships rather than the more directive techniques of psychoanalysis as the basis for developing insight.

In contrast, behavior and cognitive therapies differ in several ways from traditional psychodynamic or humanistic modes of therapy. The techniques of behavior therapies use the principles of conditioning and reinforcement to modify undesirable patient behaviors. Similarly, the techniques of cognitive therapists attempt to change feelings and behavior by changing the way a client thinks about significant life experiences.

The successful application of cognitive-behavior therapy is not dependent on patient insight and understanding. What is required, however, is patient compliance. The therapist, not the patient, is seen as responsible for the plan of treatment. Thus, patients will be told by the therapist to do things, such as writing down negative thoughts about themselves in order to come up with more realistic self-cognitions, or be trained in a system of deep muscle relaxation in order to lessen fears and tensions. To the extent patients comply with the directions of the therapist, cognitive-behavior therapists anticipate that they can help reduce and eliminate patient symptoms.

Here, then, we have two contrasting systems of therapies. Psychodynamic and humanistic therapies seek to relieve patient symptoms through interpretation and clarifications — or by encouraging the spontaneous expression of feelings — and rely upon the patients' eventual understanding of the reasons for their discontents. In contrast, cognitive-behavior therapists use techniques that do not rely on patient insights about the cognitive-affective causes of their neurotic behavior. Rather, behavior and thinking change as a consequence of the patients following the prescriptions of their therapists.

Stages in Technology

Most technologies follow a uniform path of development from human control to machine or system control. Industrial sociologist William Faunce[13] describes this evolution as proceeding from craft production in which management's dependence on employees is almost complete, to mechanized production in which dependence on labor is reduced, to automated production in which most employees are eliminated. Each stage increases the efficiency of productive activity and at the same time changes the relation between workers and machines. Thus, in their initial development, technologies are labor intensive and rely on the skills and abilities of people to ensure their successful operations. However, skilled operators are costly and even the best make mistakes from time to time. To solve these problems, technologies evolve in a predictable way. That is, they evolve by transferring the skill components of work to machines or machine systems. This transfer reduces the cost of operations, increases the efficiency of operations, and improves the quality of the products that are produced.

There are several ways in which psychodynamic therapy resembles the first stage in the development of technology. First, psychodynamic techniques are labor-intensive. By this I mean that these techniques depend on both the skills of the therapist and the capacity of the patient for insight. It is no accident that psychoanalysts write books about the "art of therapy" and the "ability to listen with a third ear." Therapeutic outcomes appear to depend upon only partially understood therapists' skills. As a consequence, there is a fair amount of uncertainty in predicting the outcomes of psychoanalytic therapy.

Another resemblance to the early stages of technological development is in terms of costs. Psychoanalytic forms of therapy frequently require 2 or 3 years to effect patient changes. As a result, the costs of treatment are high in terms of money and in terms of therapist and patient time. Because of these costs, it is generally agreed that psychodynamic therapy is difficult to adapt for mass consumption. Even if money were not an issue, there simply are not enough psychoanalytically oriented therapists available, given the length of time required for training and for treatment.

In contrast, cognitive behavior therapies represent the beginnings of the second stage in the development of a technology of psychotherapy. Many different behavior technologies have been developed and validated to help alleviate specific patient symptoms. As a result, we hear less about the art of therapy and more about techniques when cognitive behavior therapies are discussed. There is a presumption that if the therapist uses the methods that have been developed, there is a high probability that patient symptoms will be reduced. Thus, many of the artful skills needed by therapists have been transferred to the techniques of cognitive behavior treatments. A benefit of this process of deskilling the therapist's work, is that the techniques themselves can be taught in a relatively short period of time. Needless to say, like the Luddites of the early 19th century, traditional psychoanalysts have bitterly attacked what they perceive as simplified approaches to therapy.

Other benefits of cognitive-behavior therapies include a reduction in the length of treatment time, and the development of specific therapeutic techniques designed to eliminate specific symptoms. Proponents write that patient symptoms can be eliminated using cognitive-behavior techniques in as little as 4 to 6 weeks. Thus, the claim is that these newer therapies can increase the certainty of treatment at less cost to patients and in shorter time than psychodynamic therapies. Without going into the validity of these claims, which as could be expected have provoked much debate (see London,[14] for instance), these streamlining changes in therapy are consistent with advances generally described in the evolution of most technologies.

Research on Different Forms of Therapy

The simultaneous use of different technologies of psychotherapy allows us to test our ideas about behavior control techniques. That is, we would expect differences in therapists' evaluations of their patients that can be traced to the therapeutic technologies that they use.

In terms of the metamorphic model of power, I would describe cognitive-behavior techniques such as homework assignments, systematic desensitization, flooding, exposure, and hypnosis as directive and high in control (strong tactics). On the other hand, I would describe such therapist techniques as interpretation, challenging, confrontation, reflective listening, and unconditional positive regard as nondirective and low in control (weak tactics). Based on this classification, we can now make several predictions about the use of these therapeutic techniques.

The metamorphic model assumes that the use of strong tactics produces the belief that the target person's behavior is externally controlled. One prediction, then, is that therapists who use directive techniques will attribute their patients' gains in therapy to their techniques rather than to the patients' own efforts. Therapists who use nondirective techniques will attribute their patients' gains in therapy to the patients' own efforts.

The next prediction follows from a second assumption of the metamorphic model, namely, that people who are perceived as not in control of their own behavior will be evaluated unfavorably. Thus, a second prediction is that patients who are described was not responsible for their own therapeutic gains will be evaluated less favorably by their therapists than patients who are described as responsible for their own therapeutic gains.

To test the above ideas, two clinical colleagues, Finy Hansen and April Fallon, and I mailed a questionnaire to slightly over 600 practicing clinical psychologists. In the questionnaire we asked them to describe a recent case in which an adult patient in individual psychotherapy showed progress in therapy. In terms of social power, this is an instance in which the therapist has been successful in causing behavioral changes in the patient. To limit the severity of mental illness, we asked the therapist not to describe a patient diagnosed as schizophrenic, mentally retarded, organically impaired, or who was currently hospitalized or institutionalized.

We received completed questionnaires from 113 clinicians with doctoral degrees in psychology. These respondents had been in practice for an average of 14 years. Of the therapists who responded, 29% were women and 71% were men. In addition, another 35 therapists replied, but declined to participate because of such reasons as no longer having a therapy practice, not having the kinds of patients we requested, or simply not having enough time to answer.

The average age of the patients of the responding therapists was 37 years and the patients had an average of 14 years of education. There was also an imbalance in the gender of patients: 72% were women and 28% were men. Several possible explanations come to mind for this imbalance in the gender of the patients. First, more women than men may seek psychotherapy. Another possibility is that female patients are perceived by their therapists to be more easily influenced, and hence more women than men are reported as making gains in therapy.

Schools of Therapy

The clinicians represented many diverse approaches of psychotherapy. However, it was possible to group the therapists' orientations into three categories:

1. *Directive therapists* included therapists who described themselves as Cognitive, Behavioral, Cognitive-behavioral, Rational Emotive, Gestalt, or Transactional Analysis therapists. Forty-three percent of the therapists described themselves as practicing one of the above forms of therapy.
2. *Nondirective therapists* included the 29% of the therapists who described themselves as Psychodynamic, Sullivanian, Existential, Humanistic, Rogerian, Nondirective, Holistic, or Client-centered therapists.
3. *Eclectic therapists* included the 28% of the therapists who described their orientation as Eclectic or who described their orientation to therapy as including elements of both directive and nondirective therapies (e.g., cognitive-behavioral psychodynamic therapists).

Progress in Therapy

We first asked the therapists to describe the nature of their patients' progress. As could be expected, many different kinds of gains were reported including patients' reporting a greater understanding of themselves; reductions in various symptoms such as phobias, substance abuse, depression, and fears; improvements in self-confidence; and better interpersonal and family relations.

Perhaps, not surprisingly, the kind of progress made by patients was related to their therapists' orientation. Nondirective therapists reported that their patients showed increased self-understanding and improvements in their family relations. Directive therapists reported reductions in the severity of their patients' presenting symptoms. Since the goal of the cognitive-behavior school is to deal with patients' immediate problems, and the goal of nondirective schools is to increase patient insight, these differences in outcomes appear reasonable.

We were particularly interested in obtaining information about three aspects of the therapeutic relationship based on the metamorphic model. The first was the influence tactics (therapeutic techniques) that the therapist used at the time the patient made progress. The second was the therapist's explanations of the reasons for the gains made by the patient. The third was the therapist's evaluation of the patient following progress in therapy.

Therapeutic Techniques

The therapists reported using many different techniques at the time that their patients made progress. Thus, one therapist reported that he only used "silence" as his treatment of choice with his patient, presumably to force the patient to talk. Another therapist almost filled the page describing the varied interventions that she used. Here are some examples that were reported in the questionnaire.

I used a combination of nondirective reflections and feelings combined with the use of Eriksonian hypnotic techniques including time distortion, dissociation, reframing, and others.

Desensitization, cognitive restructuring,relaxation training, and response prevention.

Teaching assertiveness and getting the person to verbalize anger, use red flags to stop.

Patient listening and focused attention to her questions; support for her manifest assets; interpretation and reflection of her feelings.

Challenging the rather limited options she perceived; interpreting her current patterns in terms of repetition of her relations with parents; challenging the real meaning of her fears.

These tactics were classified into eight techniques as shown below. The proportion of therapists using each technique are given in the parentheses. These proportions add to more than 100% because most therapists reported using several techniques.

1. Psychodynamic techniques, for example, interpretation and clarification (32%).
2. Empathic techniques, for example, reflexive listening, validate patient feelings, unconditional positive regard (49%).
3. Teaching how to change behavior, for example, teaching assertive behaviors, suggestions for change, listing options, providing feedback (44%).
4. Specific behavioral techniques, for example, desensitization, breathing exercises, hypnosis, homework, (40%).
5. Cognitive restructuring techniques (38%).
6. Demands, for example, insisting that the patient comply or give up therapy (7%).
7. Therapist shared own feelings with patient (5%).
8. Coalitions, for example, third-party interventions such as bringing into therapy parents, spouses (7%).

In a preliminary analysis of these techniques, a high correlation between the use of behavioral techniques and cognitive restructuring techniques was found. Therapists who used one of these techniques also used the other. Hence, they were added together to form a new technique called "cognitive-behavior techniques," used by 51% of the therapists. We also decided to eliminate the techniques of demanding, sharing, and coalitions because so few therapists used them.

Explanations for Patient Progress

We next asked the therapists why they believed their patients made progress. We provided 11 possible reasons, each rated on a 7-point scale ranging from *of little*

TABLE 4.1. Therapist ratings of the importance of reasons for patient progress.

Reason	Mean rating	Reason	Mean rating
Relationship with therapist	(5.9)	Techniques of therapy	(4.0)
Therapist's support	(5.6)	Advice and guidance	(3.5)
Patient's inner strengths	(5.3)	Setting limits for patient	(2.8)
Patient's capacity for insight	(5.0)	Spontaneous remission	(1.3)
Therapy corrected distorted thinking	(4.8)	Pharmacological treatment	(1.0)
Therapist's interpretations	(4.6)		

importance (1) to *most important* (7). These 11 reasons and their average ratings of importance are given in Table 4.1.

As can be seen, therapists gave the highest ratings of importance to the quality of their relations with their patients, as well as to the emotional support they provided patients. Such ratings make sense because if the therapists cannot establish a trusting and supportive relationship, then it is unlikely that patients will follow the therapists' subsequent recommendations. For all therapists, then, establishing good doctor–patient relationships is the basic key to patients' progress.

This stress upon establishing relationships appears similar to the needs of physicians, until the end of the 1940s or so (see chapter 5), who also laid great stress on the need to establish good doctor–patient relations. As I will show in chapter 5, such relationship needs disappeared among physicians as medical technology provided means to cure or control most common illnesses. Perhaps as treatments for psychological ills become more certain and precise, psychotherapists may also minimize the importance of establishing supportive relationships with their patients.

Most textbooks in clinical psychology advise that the passage of time by itself can serve as a great healer of psychological problems. This means that some improvements in psychological distress may be due to the "spontaneous remission" of patient problems, rather than to therapeutic interventions. Despite this knowledge, few therapists in our sample believed that their interventions were not directly related to their patient's improvement. As can be seen from Table 4.1, "spontaneous remission" as an explanation for the therapeutic gains of patients was given a rating of 1.3, "of no importance." Therapists, then, like most humans, believe that they cause events to occur, particularly when the events are based on the use of their skills and result in positive outcomes.

The rankings of importance also mask extreme partisan variations. By this I mean that Directive therapists ranked the "'Therapeutic techniques" they used and the "Advice and guidance" they gave their patients as highly important explanations of their patient's progress. On the contrary, Nondirective therapists favored the "Therapist's interpretation" as the explanation for their patients' progress.

The techniques used by therapists were similarly related to explanations for patient progress. Users of cognitive-behavior techniques rated as more important than other therapists the "Therapist's techniques" and "Therapy corrected the patient's distorted thinking." Users of psychodynamic techniques attributed patient progress to the "Therapist's interpretations," while users of empathic techniques attributed patient progress to the patient's "Relationship with the therapist."

All of the above, of course, is simply saying that therapists who espouse a particular therapeutic orientation and use the techniques of that orientation describe their techniques as important in helping their patients. This, perhaps, documents the obvious.

Autonomy and Control

A principle concern of this research, however, was not to document the obvious, but to see whether the use of more directive techniques of influence affected therapists' beliefs about the autonomy of their patients. Two of the explanations for patient progress that are shown in Table 4.1 are based on the ability of patients to help themselves improve. These are (a) the patient's inner strength and drive toward health, and (b) the patient's capacity for insight. The therapists' answers to these two reasons were combined into the Index of Patient Autonomy. High scores meant that patients were seen as responsible for their own progress; low scores meant that they were not.

We next examined the data to see whether there were differences between therapists who used cognitive-behavior techniques and those who used dynamic techniques in their beliefs about patient autonomy. Basing expectations on the metamorphic model, we expected that the use of cognitive-behavior techniques would be associated with the perception that patients were not autonomous. Before we tested this expectation, however, patients were first equated for the severity of their psychological symptoms, because an initial analysis found that therapists' evaluations of patients were partly related to the severity of the patients' emotional problems. As a control measure, patients were divided into two groups: mild and severe. A patient was rated as presenting severe problems if the patient had a previous hospitalization for psychiatric reasons and if the therapist's diagnosis included one of these categories: major affective disorder, severe personality disorder, substance abuse not in remission, or impulse disorder. Seventy-three patients were classified as presenting mild symptoms and 40 as presenting severe symptoms.

Thirty-two percent of the therapists who used cognitive behavior techniques and 77% of the therapists who used psychodynamic techniques believed that the patients were responsible for their own progress in therapy. In terms of correlational analysis, dynamic techniques correlated positively with the autonomy index ($+ .34, p < .01$) and cognitive-behavioral techniques correlated negatively ($- .38, p < .01$) with this index.

The above findings increased our confidence in applying the metamorphic model to the behavioral sciences. Here, then, we have support for the belief that the use of strong behavioral technologies affects social science practitioners' perceptions of patients. That is, as technological control increases, self-control on the part of the patient is seen to decrease.

Evaluation of Patients

We also asked each therapist to evaluate his or her patient following the patient's progress. Evaluations were recorded based on the therapists' ratings of the following nine patient attributes: (a) Flexibility, (b) Intelligence, (c) Capacity for insight, (d) Ability to experience empathy, (e) Competence, (f) Interpersonal skills, (g) Ability to function autonomously, (h) Ability to maintain Therapeutic gains, and (i) Overall functioning.

Each of these nine attributes was rated by the therapist on a 7-point scale ranging from *very poor* (1) to *outstanding* (7). For the present analysis, these scores were simply summed to provide the Patient Evaluation Scale.

What was the relation between the measures of patient autonomy and patient evaluation? The metamorphic model predicts that those who are seen as not in control of their own behavior will be evaluated less favorably. In support of this expectation, the indices of *patient autonomy* and *patient evaluation* correlated .46 ($p < .01$). The more patients were seen as responsible for their own gains in therapy, the more favorable were the therapists' evaluations of them.

We also examined the relation between the techniques of therapy that were used and patient evaluations. Seventy-seven percent of therapists who used dynamically oriented techniques and 29% of those who used cognitive-behavior techniques rated their patients as having average or better scores on the Patient Evaluation Scale. In terms of correlational analyses, the use of psychodynamic techniques correlated positively ($+ .34$, $p < .01$) and the use of cognitive-behavior techniques correlated negatively ($- .25$, $p < .02$) with the Patient Evaluation Scale.

Clearly, then, the judgments of therapists about their patients are linked to the techniques of therapy they use. The more directive the techniques used by therapists are, the more likely therapists are to attribute patient progress to these techniques, rather than to the motivation and strength of the patients. And the more therapists believe their patients are not responsible for their own progress, the less favorable their evaluations of them.

The implications of these findings for psychotherapy need to be worked out with some care. What they strongly suggest is that as therapies are developed that use more and more specific technologies to ameliorate psychic distress, an unexpected consequence will be that therapists will denigrate their patients.

Failed Therapy

Readers may recognize that our findings are based on instances in which therapists reported improvement in patients. An interesting question to consider is what therapists may have to say about patients who they see as therapy failures. Would therapists who use cognitive-behavior techniques evaluate their failed patients more unfavorably than those using psychodynamic techniques?

Perhaps the closest we can come to an answer to this question is by looking at research carried out by organizational psychologists concerning judgments of inadequate employee performance. Extrapolating from these results suggests that therapists' evaluations may depend on their explanations for the failed therapy. If the therapist attributes failure to the inadequacy of their therapeutic techniques or to the inability of the patient to understand what is required, then the patient may be judged leniently. After all, one cannot blame the patient if the "medicine" did not work. On the other hand, if therapists have great faith in the techniques they use, then failure may be attributed to factors under the control of the patient, such as the patient's lack of motivation to continue therapy, and/or resistance to therapy. If these attributions occur, it is likely that the patient will be blamed for not showing improvement, and judged harshly. Thus, in these instances, attributions about patient autonomy also determine evaluations, but the evaluations themselves become increasingly negative as responsibility for outcome is attributed to the patient.

Some Cautions

Since the findings of this study are based on survey data, they are subject to alternate explanations. We have attempted to rule out one obvious one by equating patients for the severity of their psychological problems. However, we cannot rule out a second alternate explanation, namely, that therapists who enjoy controlling people are attracted to cognitive-behavior forms of therapy. Thus, personality differences among therapists, rather than the strength of the therapeutic techniques involved, may be the true explanation of the findings.

However, this personality explanation appears to be highly unlikely. Research literature shows that efforts to relate the personality of the therapist to the kind of therapy they choose to do have not been particularly successful. In fact, this topic was the subject of an entire issue of the journal *Psychotherapy* in 1978 (see Baron, [15] and little convincing evidence was offered one way or the other. Indeed, if such a relation were detected, it is equally plausible to suggest that the personalities of therapists may have been shaped by the kinds of therapy they practice, rather than the reverse.

Conclusion

We find, then, within a therapeutic setting what has been reported in other settings. That is, as the strength of influence tactics that are used increases, the perception that people are the autonomous source of their own actions lessens, and evaluations of them become less favorable. These findings allow us to extend conclusions concerning the control of power from interpersonal forms of influence to the use of influence based on technology. The findings support the notion that there is a strong law governing power and social relations. Simply stated, the law says that all forces that reduce individual autonomy promote devaluation of the individual. This process appears to occur regardless of the humanistic goals, kindly personality, or ethical values of the influencing agent. With regard to psychotherapy, it is reasonable to conclude that as therapeutic techniques increase in effectiveness (i.e., gain greater control over psychic distress), we can expect greater distancing between therapists and patients.

Techniques of Leadership and the Metamorphic Model

Organizational psychologists have achieved considerable understanding of the behavior of leaders in industrial and military settings over the last 40 years. One reason for this progress has been a paradigm shift in the study of leadership. Beginning in the late 1940s, applied research shifted its interest from a search for universal personality traits of effective and ineffective leaders to an interest in understanding what leaders actually do on a day-to-day basis. This shift in emphasis hit paydirt.

Psychologists at Ohio State University[16] who were responsible for this research concluded that there were two important activities carried out by all leaders. The first, called "initiating structure," or "task orientation," is concerned with the extent to which leaders directed their followers to work productively. The second set of activities, usually called "consideration" or "socio-emotional behaviors," is concerned with the extent to which leaders addressed the needs of followers for adequate communications, pleasant working environments, and the maintenance of job satisfaction. Sometimes cutting across these two components of leadership are findings about the degree to which leaders delegate decision-making powers. In this regard, autocratic leaders simply tell people what to do and democratic leaders consult and/or ask followers to decide for themselves how their work should be done.

Given the identification of these dimensions of leadership behavior, the next questions asked by researchers were whether it made a difference—particularly in terms of employee productivity—if the leader acted in a task-oriented manner or in a considerate manner, and whether people were more productive when they were told what to do or when they were consulted.

It turns out that there are no simple answers to these questions. Rather, the particular style of leadership that works best depends on the situation in which the leader finds herself or himself, and the motivations of followers. The extensive research of Fred Fiedler[17], for example, indicates that leaders should use task-oriented behavior when they have considerable formal authority, when the work of followers is routinized, and when followers admire the leader. Different strategies of leadership are advocated as variations in these situational constraints occur.

As could be expected, the research of Fiedler and of other organizational psychologists has been translated into behavior technology. Psychological consulting firms, universities, and training specialists within companies now offer courses for supervisors and managers that tell them how to diagnose the important organizational features of their work, their own leadership styles, and sometimes the motivations of followers. On the basis of these diagnoses, supervisors are trained in how to influence subordinates, sometimes using a task-oriented style, sometimes a considerate style, and sometimes a combination of both. In some training, supervisors are also coached to increase or to decrease the extent to which they delegate decision-making powers to their followers.

It seems reasonable to believe that supervisors who receive this training from behavioral specialists should view it as scientifically validated information about how to cause behavior among their subordinates. Thus, leadership becomes more than a simple "gut feeling" about how to get the best out of followers. After training, supervisors should believe that they can cause behavior using proven psychological techniques of influence.

Of course, this is just a supposition on my part. We have no information about how the use of this technology changes the beliefs of supervisors about their own efficacy. Yet, from what has been said so far about the metamorphic effects of power, it is reasonable to believe that any techniques that increase supervisors' abilities to control subordinates should also increase the supervisors' feelings of social and psychological distance from them. That is, supervisors should believe that any subsequent employee compliance to orders was caused by the supervisors' tactics of influence rather than by the employees' free choice. We would further expect supervisors to devalue a compliant subordinate's performance. This should happen because the subordinate will be given little credit for his or her own work. Rather, the work will be viewed as simply an extension of the supervisor's directions.

In theory, the seeds for industrial discord may be nurtured just a little more as a result of supervisory leadership training, not necessarily because subordinates resent being led, but because supervisors come to resent those they are leading.

I am not aware of any field studies that have examined these ideas about how psychologically trained and untrained leaders regard their followers. My colleagues and I[18] have, however, conducted a laboratory simulation of this process that provides suggestive insights. We trained university students majoring in business to supervise employees using techniques that either delegated decision-making powers or that allowed the student-supervisor to retain complete control over employees' decisions. Most commonly, these variations are referred to as "democratic" and "autocratic" forms of leadership training. A total of 678 students in many sections of an undergraduate course in organizational psychology participated.

The leadership exercise involved forming work groups consisting of one assigned student leader and five student employees. There were a total of 113 such groups. Each group was told that the exercise involved the production of paper airplanes and that the objective of the exercise was for each group to produce as many airplanes as possible.

The group leaders were then given separate instructions about the leadership styles they could use. Fifty-seven leaders were trained to act authoritatively and were given the following instructions:

You are to manage your group in a highly autocratic style. This means you are to make decisions for the group. You are to decide without consultation with individuals members how much is to be produced and who is to perform what activities . . . . Your effectiveness will be measured by your ability to maintain control over the group. (p. 326)

The instructions continued in this manner, reaffirming the leaders' need to control decision making. In terms of actual leadership training, the above instructions correspond most closely to the leadership training course devised by two organizational psychologists, Paul Hersey and Ken Blanchard[19]. In this course, leaders are instructed to simply "tell" followers what to do if followers are diagnosed by leaders as being low in "maturity."

The 56 leaders assigned to act democratically were given the following instructions:

You are to manage your group in a democratic manner. By this we mean that all decisions, such as the division of labor and the quantity to be produced are to be made by the total group. Your concern during this process should be to get the total involvement of everyone in the decision-making process. (p. 326)

These instructions correspond to Hersey and Blanchard's instructions to leaders who are directing followers diagnosed as relatively high in maturity level.

Following a period of actual work, leaders were asked to fill out an evaluation of their followers' performances. By the way, there were no differences in the actual number of airplanes produced by followers under the two leadership styles. Hence, the differences we found in these evaluations are not due to objective performance differences that resulted from the two leadership approaches.

There were three sets of information gathered. The first asked whether authoritarian and democratic leaders differed in the kinds of tactics they used to influence group members. We asked leaders and group members, independently of each other, to provide this information. Autocratic leaders were described by followers, and described themselves, as using the following tactics: "Simply order people to do as asked," "Set time deadlines," and "Demanded that people do as they were told." On the contrary, leaders trained in democratic forms of leadership were described by followers, and described themselves, as "Using mutual discussions," "Letting people work on their own," and "Asking in a polite manner."

The second set of information asked the leader to estimate the extent to which they believed their subordinates' performances were due to the subordinates' own motivations to do good work. The third set of information asked leaders how satisfied they were with their subordinates' performances.

The statistical procedure of path analysis was used to examine this information. Based on this analysis, the following explanations of the results were offered. Leaders who were trained to act either autocratically or democratically used controlling or noncontrolling influence tactics respectively. In turn, the use of either of these sets of tactics differentially changed the leader's belief about the subordinates' autonomy. That is, leaders who used controlling influence tactics reported that their subordinates were not self-motivated. On the other hand, the leaders who used noncontrolling tactics said that their subordinates were motivated to do good work.

Also, the path analysis supported the assertion that the major determinant of favorable employee evaluations was the leader's assumption about the extent to which his or her subordinates were self-motivated. Subordinates who were seen as self-motivated were evaluated more favorably than were those not so seen.

Finally, the path analysis revealed that there was no direct link between a subordinates' actual performance and the leader's evaluation of that subordinate. That is, if two employees were doing equally good work, a more favorable evaluation would be given to the employee who was seen as self-motivated and a less favorable evaluation would be given to the employee seen as forced to work by the supervisor's orders.

The results of this study are exactly equivalent to the routinization of work studies that will be reported in chapter 6. In chapter 6, however, control of the pace and quality of employee performance was based upon a mechanical technology (i.e., conveyor belts). In the present research, control was based upon a behavioral technology (i.e., teaching leadership styles). Nevertheless both sources of control transformed the supervisors' thinking about the efficacy of their employees in identical ways. The more control they exercised, the less favorable the evaluation of their employees.

Conclusion

The major conclusions can be summarized in Figure 4.1, which traces the flow of events that produce different experiences among therapists and supervisors. Let us follow the sequence of the more common events. Following the arrows for the cognitive behavior therapists, we see that the techniques they use are straightforward sets of directions and exercises for patients. If patients comply and follow the therapist's regimen, symptom reductions occur. For the therapist, the most likely explanation for these reductions in symptoms is that their techniques were effective. Little credit for improvement is given to the patient, who is simply expected to comply. Unfortunately, one consequence of denying patient responsibility is that the patient is given unfavorable evaluations.

Similarly, if we follow the arrows for supervisors instructed to use autocratic forms of leadership we find the same pattern of outcomes. If employees accept the leader's orders and perform well, the leader assumes that his or her orders caused these outcomes. Again, the perception that employees are controlled by outer rather than inner forces leads to less favorable performance evaluations.

Behavior Technology		Influence Techniques		Explanation of Progress		Evaluation of Person
Cognitive-behavior therapy	→	Relaxation Homework Desensitize etc.	→	Techniques caused improvement	→	Less positive regard
Autocratic leadership training	→	Strong verbal tactics	→	Forced to work	→	Negative performance evaluations
***						
Psycho-dynamic therapy	→	Interpretation Feedback Discussion	→	Patient insight caused improvement	→	Responsible, positive regard
Democratic leadership training	→	Rational & weak tactics	→	Employees own motivations to perform well	→	Favorable performance evaluations

FIGURE 4.1. Flow of events linking behavior technology and evaluations of persons being influenced.

If we look at the bottom set of arrows we see that the influence techniques of the psychodynamic therapist involve discussion, interpretation, feedback, and confrontation. Patient compliance, in this instance, involves the patient coming to understand the basis for his or her problems, and then using these insights to change thinking and behavior. The consequence here is that therapists hold their patients in high regard and evaluate them favorably.

A similar interpretation is offered for supervisors who were trained to use democratic forms of leadership. The perception that an employee is voluntarily complying is sufficient to increase the supervisor's attraction to that employee, and to cause the supervisor to evaluate him or her favorably.

While these interpretations are consistent with the theoretical perspective of the metamorphic model, I recognize that the effects of behavioral technologies on users may be trivial. It is unclear, for instance, whether the negative evaluations made by cognitive-behavior therapists of their patients are cause for any worry. Indeed, for several decades, cognitive-behavior therapists have minimized the importance of the therapist–patient relationship. It is argued by therapists of this theoretical persuasion that gains in therapy are contingent upon patient compliance rather than upon whether or not the therapist evaluates the patient positively. From this perspective, the analogue is to medical practice, where it is not necessary for physicians to like their patients in order for the patients to get well. Rather, improvement is contingent on the physicians' prescriptions of the correct medicines and treatments.

If we want to know whether the use of behavior technologies promotes exploitation among their users, we must examine its use in many different settings. We simply do not know at this time how the successful application of behavior technology to change, for instance, people's attitudes or to program their learning may change the influencing agent's attitudes and values.

In settings such as therapy, for instance, technology cannot tip the balance of power too far in favor of the therapist simply because patients also control resources that therapists want (i.e., money). However, when recipients control few resources of value, or when recipients are not aware that they are being subjected to scientifically validated forms of influence, then the added control provided by behavior technology may seriously disrupt social relations and produce exploitive behaviors.

Thus, much remains to be done. The evidence presented in this chapter demonstrates that the metamorphic model can provide insights and findings about the use of power that, until now, have not been considered by psychologists. Further studies of these effects in the many different areas of applied psychology are needed before definite conclusions can be reached.

References

1. London, P. (1969). *Behavior control*. New York: Harper & Row.
2. Clark, H.B. (1987). *Altering behavior: The ethics of controlled experience*. Beverly Hills, CA: Sage.
3. Cartwright, D. (1965). Influence, leadership, and control. In J. March (Ed.), *Handbook of organizations* (pp. 1-47). Chicago: Rand-McNally.
4. Cialdini, R.B. (1985). *Influence*. Glenview, IL: Scott Foresman.
5. Freedman, A.E., & Freedman, P.E. (1975). *The Psychology of political control*. New York: St. Martin's Press.
6. Halleck, S.L. (1971). *The politics of therapy*. New York: Science House.
7. Gartrell, N., Herman, J., Olarte, S., Feldstein, M., & Localio, R. (1986). Psychiatrist-patient sexual contact: Results of a national survey. *American Journal of Psychiatry, 143*, 1126-1131.
8. Back, K.W. (1972). *Beyond words: The story of sensitivity training*. New York: Russell Sage Foundation.
9. Lieberman, M., Yalon, I., & Miles, M. (1971). The group experience project: A comparison of ten encounter technologies. In L. Blank, G. Gottsegen, & M. Gottsegen (Eds.), *Encounter, confrontation of self and interpersonal awareness*. New York: Macmillan.
10. Kaplan, R.E. (1983). The perils of intensive management training and how to avoid them. *Professional Psychology, 14*, 756-770.
11. Guy, J.D., & Laboie, F. (1986). The impact of conducting psychotherapy on the psychotherapist's interpersonal functioning. *Professional Psychology, 17*, 111-114.
12. Dutton, D.G. (1973). Attribution of cause for opinion change and liking for audience members. *Journal of Personality and Social Psychology, 26*, 208-216.
13. Faunce, W.A. (1981). *Problems of an industrial society*. New York: McGraw-Hill.
14. London, P. (1986). *The modes and morals of psychotherapy* (2nd ed.). New York: Hemisphere.
15. Baron, J.B. (1978). Preface: The theory and personality of the psychotherapist. *Psychotherapy, 15*, (special issue) 307.

16. Fleishman, E.A., Harris, E.F., & Burtt, H.E. (1955). *Leadership and supervision in industry.* Columbus: Ohio State University, Bureau of Educational Research.
17. Fiedler, F. (1967). *A theory of leadership effectiveness.* New York: McGraw-Hill.
18. Kipnis, D., Schmidt, S., Price, K., & Stitt, C. (1981). Why do I like thee? Is it your performance or my orders? *Journal of Applied Psychology, 66,* 324–327.
19. Hersey, P., & Blanchard, K. (1982). *Management of organizational behavior.* Englewood, NJ: Prentice-Hall.

CHAPTER 5

Medical Technology

Over the last 200 years medical technology has transformed the practice of medicine.[1] The fumbling, inept, 18th-century physician has been replaced by assured teams of physicians, often operating their own corporate centers of medicine. Family medicine is now as likely to be practiced by board certified experts employed by corporate medical centers as by general practitioners using their own homes as offices. Discoveries in medicine have provided the physician with a range of strategies for controlling disease. Physicians now, as never before, have the final say concerning questions of health, and most patients readily agree to do whatever the physician recommends. It is apparent that medical discoveries have not only shifted the ways in which medicine is practiced, but also shifted the balance of power between patients and physicians in favor of physicians. This chapter is about these changes in social power and how medical technology caused them to occur.

From the early 1900s until the 1970s, many surveys suggested that family physicians were highly regarded by Americans (see Starr[1]). One reason for this esteem was that physicians were seen as concerned, caring, and self-sacrificing of their personal lives in order to minister to their patients. One wonders why other professionals, such as accountants and lawyers, were not seen through the same humanistic screen as physicians? Surely as individuals they cared for their clients as much as physicians.

I believe that there are several explanations for the flattering portrait of physicians. The first originates in the belief that people who help us must be kind. That is, if we believe in a just world, it follows that we must also believe in a compassionate physician. When we are ill and helpless, we surely would not want to believe that an unkind person, or worse, is taking care of us. We are reluctant to abandon the image of the comforting personal physician as we are to abandon the image of Santa Claus. Who would willingly suspend belief in the possibility of competent, caring kindness, especially when one is ill?

A second explanation is that the early 20th-century physicians could actually cure many illnesses. The wondrous discoveries of the 18th and 19th centuries finally became available as practical technology for everyday use with patients. Thus, the image of physicians was based in part upon their ability to restore health. In explaining the personal prestige of the doctor of the 1920s and 1930s, Shorter,[2] for instance, wrote that people were willing to submit to the doctor because they saw him as a progressive representative of science.

A third explanation is that the stereotype of self-sacrificing physicians was based upon their actual behavior. That is, physicians as a group behaved in ways that were consistent with this compassionate image. I believe the reason for this service orientation, quite simply, was that physicians lacked power. Perceiving themselves as less than powerful meant that physicians had to present a kindly face to the world. If they did not "act nice," it was likely that they would not build their practice, and hence would not survive economically.

Unfortunately, the image of the kindly physician portrayed in advertisements, nostalgic television programs, and in the minds of many, is disappearing. Gradually emerging is a new image of the physician as a medical technocrat and impersonal dispenser of health care. Times have changed and so, too, has the practice of medicine. Corporate medicine, specialists, group practices, hospital care, and even newer forms of health care now dominate the practice of medicine. Medical insurance companies that provide third-party payments are increasingly limiting the physician's ability to prescribe hospitalization and/or expensive medications.

In what ways are these changing images a reflection of newer discoveries in medicine? The thesis of this book is that technology alters social relations by altering the balance of power between persons. Thus, I will argue in this chapter that to understand the changing image of the physician, we must look to technological innovations in medicine. That is, as technology has reduced uncertainty in treating illness, there have been concomitant shifts in the balance of power between patient and physician, as well as accompanying shifts in social relations between the two. These shifts have reduced patients' power and increased physicians'.

The Balance of Power in Doctor-Patient Relations

In chapter 2, it was pointed out that the balance of power between people can be described in terms of two sets of events. The first is the extent to which each person wants something that requires the services of others. The second is the extent to which each person has the means or resources to persuade others to satisfy their wants. Where the needs of both parties are of equal strength and each has the resources to help the other out, the balance of power between the two parties is relatively equal. In terms of this equation, the resources and needs of patients and physicians are fairly obvious.

Patients' needs are often extreme and urgent. They are sick and wish to be healthy. The sicker the patient, the more desperate this need. Patients fear cancer or a failing heart. In truth, most patients wish to live forever, not just for themselves, but for their families, and for those who would cry if they died. Thus, sick people will go anywhere, do anything, let anything be done to them, and most important, willingly use all of their financial and material resources in exchange for treatment.

To illustrate, in Johnstown, Pennsylvania, in the late 1950s many citizens supported a self-styled doctor's claim that he could cure cancer. I was sent to

Johnstown by the American Cancer Society to see if I could find out the basis for the doctor's popular support. The cure, I was told, was based upon a salve discovered by the "doctor's" father in Texas. Originally, the salve had been used, he said, to cure cancer in horses. While the physicians of Johnstown were united in their opposition to such quackery, the ordinary citizens were more tolerant. In fact, a few weeks before my visit, there had been an attempt to set fire to the Johnstown hospital. Some citizens were actively protesting what they believed to be physicians' attempts to suppress a cure for cancer. The belief was fairly common that physicians made money from treating cancer, even if they could not cure it.

During this time, patients who suspected they had cancer flocked to the treatment tent on the outskirts of town. Why did they go? In interviews with them and with their relatives, I found that many persons with cancer were treated at hospitals and then discharged. The large number of surgical, chemical, and radioactive treatments that exist today were unavailable at that time in Johnstown. Consequently, patients were often told that there was nothing else to be done. For all practical purposes, they were told to go home and wait for death. Not surprisingly, most of these patients refused to give up. They flocked to the "doctor's" tent. Happily, they paid from $10 to $20 per treatment for the wondrous salve that might prolong their lives.

The dependency of patients, then, is unending. From the perspective of social power, the wish to be well forces patients into willing submission to anyone who claims, truly or not, the ability to relieve their distress. What do physicians want from patients? Clearly they have many wants that involve an exchange between themselves and their patients. At the highest level, physicians want the satisfaction of helping others, of gaining the reputation for competence, and being of service to their community. But the exchange of fees for services is the "carrot" that attracts physicians to patients. That is, physicians' economic motives must first be satisfied by patients before most physicians attempt to satisfy higher order needs. Particularly in the United States, the model of the individual entrepreneur physician is the one most frequently supported by medical associations and individual doctors. It seems safe to say that few, if any, physicians would work without payment for very long.

In short, illness and treatment form a perfect basis for the development of power relationships. Both parties have needs and both parties have resources needed by the other. As these resources create mutual dependencies, power relations between the two parties approach equality. As the physician becomes more dependent on the patient's resources than the patient on the physician's, or vice versa, then power relations favor the least dependent party.

Pre-World War II

With the analysis of needs and resources as background, I can now attempt to answer why the image of the caring, self-sacrificing physician has so long existed in this country. Let me begin with a description of the stereotype of physicians as they might have been seen from, say, the 1920s to the 1940s.

First, it would not have been unusual for the physician of that time to be privy to family problems, and in all ways act as confidant and advisor of the family. The general expectation was that the physician would be available day or night, ready to sacrifice a Sunday's rest or a late night's sleep to treat illness. Few persons would have hesitated to disturb the Sunday of a physician if someone in the family was sick and if they could pay for the physician's services. To put these remarkable expectations for service in perspective, by contrast, few persons then or today would think of calling their accountants or lawyers on Sunday and request a home consultation about problems of money or law. Clearly the self-sacrificing role of physicians was uniquely established as a social norm during this time period and earlier.

To move the description along, we now witness the physician arriving, small black bag in hand, smiling but concerned, even apologetic for delays that were not of his causing, but results of his duty to others. Perhaps if the emergency was not too great, there even would be time for chitchat about past happenings in the family – marriages, illnesses, children – that showed the physician was truly a family advisor and friend.

Then the ritual would begin. First consultation – who was sick, how long, and the symptoms – then the trip upstairs, the scrupulous scrubbing of hands at the beginning, the unhurried questioning about symptoms and whatever else the patient might want to mention. Then came the examination: thermometer into the mouth; hands on the throat and the abdomen; taking the pulse; tapping the chest and back; using the stethoscope to listen to the heart and the lungs; reading the temperature; a final peeping into the mouth, the throat, the eyes, and the ears; and then the examination was complete. Through a good bedside manner, the physician had obtained a medical history and completed a thorough, modern 1930s examination. All that was left was the oh-so-careful washing away of bad health, the diagnosis, the walk back down the stairs, and the advice. The patient got better or not, as the case may have been. Better or worse, there was little cause for complaint. The second opinion was obtained from a knowledgeable cousin or aunt, but surely not from a second physician. To do so was to destroy the trust and confidence that, in fact, gave rise to healing and health.

Why should physicians of 100 years ago, and even of 40 years ago, have spent their time chitchatting and acting nice? Why should they have sacrificed Sundays with their own families to wait on others? None but saints would freely choose to live at the call of others. Yet physicians throughout the 1940s did so, but they do not do so today. Surely the human nature of physicians has not changed, in this short period, from humanist to busy technician.

Why Physicians "Acted Nice"

As I suggested earlier, one way to answer the above questions is to look at physicians' power in relation to their patients'. In earlier chapters, it was pointed out that social relations are guided in part by the balance of power that exists between persons. The greatest civility in social relations exists when this balance is relatively equal, that is, when both parties have needs that the other can satisfy. However, when one person wants much from the other but has little to give in return, the balance of power is disturbed. Under these circumstances, as I described in Chapter 2, the less powerful party frequently acts in an ingratiating, pleasant way, calculated to give little or no offense. Both in interpersonal relations and in work settings, people with less power try to get their way by presenting an image of concern, self-sacrifice, and helpfulness. From this perspective, the kindly behavior of the physician, designed to be of service and to please, may have been indicative of powerlessness rather than humanitarian impulses.

The next section of this chapter will describe how the balance of power favored the patient rather than the physician throughout the 19th century, and gradually approached equality by the midpoint of the 20th century. The fact that physicians were still service oriented throughout the 1950s and beyond suggests that the physicians' definition of appropriate behavior for themselves, established in earlier years, changed at a slower rate than the physicians' actual power. But throughout the 19th century, the physician's dependence on the patient was greater than the patient's on the physician.

Among the reasons for the physician's dependence prior to the 20th century were (a) the physician's uncertain ability to cure: (b) the intense competition for patients from nonmedical health providers; (c) the populist tradition of being able to cure one's own illnesses – a tradition that still persists in this country; (d) the fact that most physicians were general practitioners who depended on repeat business to build up their practices; and (e) the need of physicians to maintain good rapport with patients in order to diagnose the patients' illnesses.

Inability to Cure Disease

Perhaps the foremost limit was the fact that physicians had no real means to cure many of the illnesses that they encountered. In the 18th century and throughout most of the 19th century, treatment was limited to such mindless techniques as bleeding, vomiting, blistering, and purging. Diagnoses were uncertain and usually not related to the actual causes of illness.

Simple medicines, herbal treatment, and the like set limits on physicians' actual ability to cure. The harm done to many patients as a result of bleeding, purging, and other deadly forms of treatment, produced a consensus among informed physicians that it was perhaps better not to treat at all. In 1755, Boston physician William Douglass wrote: "excepting in surgery and some very acute cases, it is better to let nature take her course" (quoted in Shorter,[2] p. 20).

This lack of real knowledge set the stage for physicians' public behavior. At a minimum, physicians had to present an image of expertise. In his award winning book, *The Social Transformation of American Medicine*, Paul Starr[1] describes the importance of self-presentation techniques for physicians throughout the 19th and early 20th centuries. Physicians had to sell themselves because they had little effective technology to sell. Perhaps recalling Thomas Hobbes's dictum that "the reputation of power is power," physicians had to create an appearance of wisdom.

Starr describes a popular manual for the medical practitioner published in the late 19th and early 20th century, called *The Physician Himself*, by D.W. Cathell. This author advised physicians to concern themselves first with appearing competent, and only second with actually being competent.

Well into the 20th century, the family physician was limited in his ability to control illness. Family diseases were allowed to run their course, with aspirin, quarantines, and knowledgeable advice on the probable length and course of the disease. This was the best that the physician could offer. As Marvin Edwards[3] points out, the doctor's main assets were education and a comforting bedside manner. His tools were limited and primitive and the pharmaceutical armamentarium was meager at best. Antibiotics and other life-saving medications were unknown just a few decades ago. Surgical techniques for the treatment of cancer, coronary heart disease, and other major causes of death had not been developed.

In short, professional knowledge and expertise were limited up through the 1930s. Patients, of course, recognized that in many instances physicians could do little beyond diagnosing and providing information. Expectations for cure were far lower than those that exist today. But recognition that there was little a physician could do reduced the physician's value to the patient, and hence the physician's potential power.

Competition

A second limit on the physician's power throughout the 19th century was competition, which allowed patients to choose from a variety of potential health care givers. There were herbalists, botanists, medicine men, and out-and-out quacks who called themselves physicians. A second variety of alternative physician was the local folk healer: wisewomen, midwives, bonesetters, and numerous other sorts of paramedicals, who for centuries had provided basic medical care to small towns and villages. In short, scientific medicine inspired so little faith that everywhere rivals arose to cut deeply into the physician's trade.[2]

A further source of competition was from the physicians' ranks themselves. Medical mills of the 19th century produced an unending supply of people with paid-for medical degrees, if not medical knowledge. By 1860, there were 47 medical schools, which had turned out about 17,000 graduates. Most were little, proprietary schools, founded by a few doctors who lived from the student fees. "There were no laboratories, little dissection, and few chances to see patients. For the most part, graduates were absolutely incompetent to assume the duties of practitioner[2]" (p. 143). Needless to say, this stopped few from actively entering the medical profession.

The simple result of this competition was that physicians had to try harder if they were to earn a livelihood. A refusal of service simply meant that the patient would turn to one of the dozens of other purveyors of health care that were available. Pharmacists and patent medicines offered as ready a promise of cure for illness as physicians. Credentials from graduates of diploma mill colleges were as readily accepted by state licensing regulations as the diplomas and credentials from established medical schools, such as Columbia University and Johns Hopkins University.

The Mental Health Profession

Perhaps we can better understand the uncertain competitive position of the physician prior to the 1930s by examining the struggles of today's mental health practitioners to limit competitions. Today, ministers, social workers, psychologists, group dynamicists, and more are all in fierce economic competition for patients. Some practitioners have a doctoral degree, others have a master's degree, and still others have less formal education. Some mental health practitioners have been trained in schools licensed by social work associations, others by psychological associations, while still other practitioners have taken courses in therapy from nonlicensed schools, or even through correspondence courses.

Thus, the person who is experiencing symptoms of mental ill health has a bewilderingly large number of persons who are willing to offer treatment in exchange for money. The patient has only to decide which particular kind of therapist he or she wishes to hire. Behaviorists, eclectics, dynamic Freudians, Gestaltists, humanistic therapists, psychodramatists, and more all compete for the client's patronage. And, if the client is dissatisfied he or she can readily find another therapist. Such ready access to service leaves most mental health practitioners cultivating the earlier style of the physician. That is, winning loyalty by presenting themselves first as competent, expert, and professional, and second as concerned and caring for the client.

A further parallel between the experience of mental health practitioners of today and the medical physicians in the 19th century concerns the ability to cure patients. The 19th-century physicians could not demonstrate that their treatment was effective, and today's mental health practitioners are experiencing similar problems. This lack of ability to demonstrate competence has been a major reason for the reluctance of many state legislatures to pass licensing laws. For physicians, effective licensing laws were passed only after the discoveries of the 19th century that led to the cure of many germ-based diseases. Such cures demonstrated to licensing boards that medically trained physicians possessed unique qualifications. Once passed, these licensing laws helped reduce competition from nonmedically trained practitioners.

The mental health movement awaits similar dramatic discoveries. Until then, however, mental health practitioners will remain in oversupply, and consequently will experience the same need to manipulate their public images as the 19th-century physician.

Self-Care Movements

Another limit on the early physician's power originated in the belief that people could and should be responsible for their own health. This populist belief persists to this day, as shown by recent books such as *Women's Guide To Their Own Bodies*. Rather than consulting the doctor, many families consulted guides, almanacs, and other sources that listed simple remedies and home cures. Written in everyday language, the books set forth current knowledge on diseases and how to treat these diseases.

Another factor contributing to self-care in the United States throughout the 19th century was the fact that many families lived far away from medical care. It was difficult to travel 40 miles or more over poorly tended roads and trails to obtain medical help. As a consequence, the family was the center of social and economic life and the natural focus of care of the sick. Women were expected to deal with illness in the home and to keep a stock of remedies on hand.[1]

In short, both the populist theme and geographic factors served to reduce the unique value of physicians during the 19th century.

It is of some interest that as the medical profession gained in status and reputation during the first three decades of this century, it vigorously combated the idea that people could act as their own physicians. Clearly, the medical professions grasped the idea that in order to increase one's power, it is necessary to make the other party dependent on what you have to offer. Advertisements in popular family magazines stressed the theme that mothers were unable to recognize the severity of symptoms in their families or to cure a sick family member. Always, the message included a stern warning to call the doctor at the first sign of illness.

Repeat Business

A fourth factor limiting the power of the physician originated in the physician's need to maintain the loyalty of patients. Before the 1940s, roughly 75% of all physicians were general practitioners. This meant that they were the family doctors, consulted on each occasion of illness. As an independent practitioner, the physician had to provide the money for medical equipment, rent, and furnishings. The choice of a bad location, the perception by patients that the physician had a "bad" personality (cold, alcoholic, unfriendly, frivolous), or rumors of technical incompetence were usually enough to force the physician out of business. Success, on the other hand, depended on repeat business from the same families. Physicians, then, had to act nice if they were to build a steady practice. These economic forces associated with repeat business increased physicians' dependence on their patients' goodwill, and consequently added to the physicians' need to project a kindly and competent image.

Bedside Manners

Until the 1950s, one of the more important criteria for judging the worth of a physician was his ability to diagnose the causes of illness. As Shorter[2] wrote, the point of medical training was to produce doctors who would have some sense of what was going on when a patient came in with a pain in the chest or blood in the urine. This meant that the doctor needed the patient's cooperation to uncover significant medical history. An incorrect report of symptoms could lead to an incorrect diagnosis.

From the late 1870s up through the 1950s, technology provided physicians with many instruments to help this process of diagnosis. Devices became available that helped to listen to the heart, to check the flow of blood, and to examine urine with a microscope. X-rays became available to look inside the body.[1,4] Yet throughout this time period, the diagnosis was made on the basis of the physician's integration of information from the patient and from the diagnostic instruments. The more brilliant the physician, the greater the ability to match the patient's medical history and physical examination with known causes of disease.

The physician's bedside manner, then, contributed strongly to the diagnostic skills of the physician. Throughout the 19th century, the physician was almost completely dependent on the patient's retelling of the symptoms. If the physician was brusk, hostile, or impatient, it was possible that the necessary information would not be offered. Even until the mid-20th century, it was the physician who was expected to integrate information from interviews, inspection of the body, and laboratory tests when diagnosing illness. His or her skill in interpreting signs and symptoms formed the basis for diagnosis. A wrong diagnosis led to the wrong recommendations for treatment, and a patient's verbal reports were a significant core of the diagnostic procedure. As a consequence, physicians had to act in such a way as not to intimidate or threaten the patient. This need, then, further contributed to the physician's pleasantness.

Modern Technology and the New Physician

Through the 1950s, then, the physician's lack of power helped shape his relations with patients. The inability to cure, competition, the dependence on the patient for repeat business and for diagnostic information all contributed to the physician's attempts to maintain cordial relations. Uncertainty in treatment, uncertainty in business, and uncertainty in diagnosis all led to the image of the compassion and self-sacrifice of the physician of the pre-World War II era.

Important changes have occurred since the 1950s in the practice of medicine. For instance, among American physicians in 1928, 74% were general practitioners. By 1980, however, of the 403,000 physicians in the United States, only 15% described themselves as general practitioners.[1]

In direct proportion to the decline in the number of family physicians has been the decline in home visits by physicians. The house call as common medical practice has disappeared. From July 1966 through June 1967, only 3.4% of all physician visits with white patients took place inside the patient's home. The figure was even lower (2.2%) for nonwhite patients.[3]

Most physicians would argue that there is no longer any need for home visits. No longer can the physician cram into his black bag all the equipment he will need for his practice. As Edwards[3] writes:

The modern physician's office is a storehouse of highly developed diagnostic and testing equipment. His office has become a miniature hospital. . . . The truth is, any doctor who routinely rushed to a patient's home with the medical tools he could fit into a tiny black bag would be guilty today of first rate medical incompetence—and probably would be rewarded with poor results. . . . Today the patient can reach the doctor's office as quickly as the doctor can reach the patient. (p. 47)

While such arguments seem reasonable, they overlook the fact that this shift in locus of service also reflects a subtle shift in power. Social-psychological and ethological studies find that people are more assertive and controlling in their own territories than when outside. By extension, one can expect patients to be more submissive when waiting in the doctor's office than when speaking to the doctor in their own home. This shift in geography gives the physician home court advantage. You must wait for the doctor to see you rather than vice versa.

Just as there have been shifts in the locus of treatment, there have been shifts in social relations between physicians and patients. Doctors have seemingly lost interest in their patients. Descriptions today of physicians are less likely to mention warmth and compassion. Rather, the emphasis is on competence and impersonality. Physicians have joined the accountant and the lawyer insofar as their public image exists.

How can we account for this almost overnight change in social relations? The answer is that physicians today have far more power than their counterparts of the early 1900s and their behavior simply reflects this newly found power. They do not have to act nice or cater to the needs of patients. If anything, patients must cater to the physicians' needs.

Many factors have contributed to this shift in the power of the physician vis-à-vis patients. Foremost are the medical discoveries that allow the physician to control disease with greater precision than in the 1930s.

The Reduction of Uncertainty in Treatment

By far the most important reason for patients' increasing dependency on the physician has been the increasingly accurate means of dealing with illness. This base of power emerged slowly in the late 19th century with the discoveries of Koch and Pasteur, among others, of the role of germs in causing illness. At the same time, chemistry and biochemistry were providing physicians with drugs and medicines that truly helped the patient. There were drugs to cure pain (such as morphine, aspirin, and codeine), drugs to sedate the patient (such as chloral hydrate), and barbiturates. Ether and chloroform were first used in the 1840s, quinine was introduced in the mid-19th century, antitoxins against diphtheria were developed in the 1890s and used to inoculate children, failing hearts were helped by the use of digitalis, and a cure for syphilis was developed. Such discoveries provided physicians with the means to cure diseases that had formerly been almost inevitably fatal.[1,4]

The modern period of medicine began in 1945, with the use of penicillin. In the next decade, a whole series of antibiotics became available, effective mainly against bacterial infection. These wonder drugs transformed our encounters with disease; such illnesses as pneumonia, gonorrhea, and rheumatic fevers all but disappeared. Other drugs preventing inflammation in the joints, making the arteries larger in fighting high blood pressure, causing the heart to beat more slowly or rapidly, composing the mind, and thinning the blood were discovered and used beginning in the 1960s. By the 1980s, entire new sciences such as immunology and clinical biochemistry had increased the ability of physicians to cure disease.

Keeping pace with biochemical discoveries has been the development of operative techniques undreamed of in the 1940s. Physicians are now able to replace parts of the body. The speed of medical discoveries suggests that, tomorrow, gene splicing and further understanding of the biochemical basis of aging will lead even old age to be labeled a disease rather than a natural process of life.

One final point to be mentioned concerns the ability of physicians to diagnose causes of patient illness. In the 1980s, the art of diagnosis has been transformed by biochemistry and by machines that can examine directly the interior of the heart, the brain, the stomach, and the lungs. No longer does the physician have to spend long hours talking to patients to get their medical histories, or to piece together from signs and symptoms an accurate diagnosis. Medical technology has taken this art away from the physician. Today, precise laboratory reports provide far more accurate diagnoses than the physician's guesses. The physician, then, is freed of the need to adopt a bedside manner; pleasant attitudes toward the patient no longer help determine the accuracy of the diagnosis.

In short, medical technology has provided physicians with many techniques that reduce uncertainty in treating illness. Antibiotics, new drugs, and new diagnostic and operative procedures allow physicians to control disease with far greater certainty than yesterday's home visits, family chitchat, tea, aspirin, and sympathy.

Scientific Knowledge Replaces Ordinary Knowledge

Another source of increased power for physicians has been the change in public attitudes toward physicians. By the early 20th century, most persons believed that physicians possessed knowledge that was no longer available to the ordinary lay person.[5]

We recall that the ability to cure disease was possessed by many persons besides doctors. Throughout the 19th century and before, medicine's unique claim to specialized knowledge seemed to be based as much on the use of language not commonly available to lay persons as upon the ability to heal. The body and diseases were described in Latin or Greek, or with jawbreaking names. If this use of language did not cure, it reduced uncertainty by naming what was wrong. Needless to say, the amount of respect and power accorded this kind of knowledge was frequently viewed with bemused tolerance, rather than respect, by intelligent lay persons of that time. Often, the pompous language of medicine was seen by comics of the 19th century as a source of humor.

These attitudes changed rapidly as medical science developed proven cures for common illnesses. Starting in the 20th century, scientific reports, the media, and people's own experiences convinced the public that physicians controlled real knowledge about health that was unavailable to nonmedical persons.

Illich (1976) summarizes the enormous growth in power given to physicians by public opinion in his book, *Medical Nemesis*[5]:

Only doctors now know what constitutes sickness, who is sick, and what shall be done to the sick. In some industrial societies, social labeling has been medicalized to the point where all deviance has to have a medical label. The eclipse of the explicit moral component in medical diagnosis in favor of science, has thus invested medical authority with totalitarian power.... In a medicalized society, the influence of physicians extends not only to the purse and the medicine chest, but also to the categories to which people are assigned. Medical bureaucracies subdivide people into those who may drive a car, those who must stay away from work, those who must be locked up, those who may become soldiers, those who are competent to commit a crime, and those who are liable to commit one. (p. 47)

Restriction in Competition

Beginning in the early 1900s, the American Medical Association established a policy of evaluating and grading the quality of medical school training. One consequence of these new standards was that the number of medical schools in the country dropped from 162 to 95 during the years 1906 to 1916. Standardized curricula were established that lengthened medical training to 3 years and more. Rather than the minimal training of as little as 6 months offered by some schools, medical students soon were faced with academic requirements that meant they might not begin to practice medicine until the age of 30. In addition, the number of students admitted to medical schools was drastically reduced. Harvard, for instance, in 1914 limited its medical school class admissions to 125 from well over 200 in earlier years.[1]

These new standards resulted in physicians becoming scarce commodities in many parts of the country. Thus, at precisely the time when patients were accepting the idea that medicine made a difference, physicians' availability diminished. Since power is enhanced by scarcity, the reduction in competition caused by tougher standards clearly served to increase physicians' power in relation to their patients, the sick had to compete for the services of fewer doctors.[6]

Specialization of Medical Practice

The next set of events that has reduced patients' power is the rise of medical specialization. As I mentioned earlier, before World War II, over three quarters of physicians in active practice reported themselves as general practitioners. The percent reporting themselves as medical specialists jumped from 24% in 1940, to 37% in 1949, to 44% in 1955, to 55% in 1961, to 69% in 1966.[1]

A major change has occurred, then, in the way in which people receive medical care. It is less likely today that patients can walk into a doctor's office and expect treatment. Symptoms and specialization must first match, or the patient will be referred elsewhere.

Why have physicians specialized? Why haven't they continued to treat the whole person, regardless of the site of illness? The answer, of course, rests in the rapid speed with which medical knowledge has been converted to application. The generalist who treats all diseases simply cannot keep up with discoveries, developments, and new treatments. The recent histories of accounting, law, psychology, and biology illustrate the same process of fragmentation as in medicine. Knowledge-based professions are inevitably driven to focus more and more intensely on less and less.

Furthermore, the income of medical specialists has grown at a faster rate than that of the general practitioners. For these reasons and more, the family physician who knew all about the patient has been replaced by specialists who know about perhaps only one of the patient's organs. A consequence of this process of specialization has been a further weakening of the economic dependence of physicians upon their patients.

This has happened because by reputation the medical specialist can heal what the general practitioner cannot. Since patients have already heard their own physician say the dreaded words, "I cannot help you. Go see a specialist," their needs are major. The judgment of their physician also confirms that the specialist is their last resort. For these reasons, the specialist has achieved a position where he or she can charge more money and the patient cannot refuse. How can one haggle about price when there is only the specialist to help?

Specialization frees physicians from dependence upon patients in another way as well. Most specialists, except for perhaps pediatricians and gynecologists, do not depend upon repeat business. The expectation is that the patient will consult about a particular problem and may never be seen again. Hence, specialists do not have to act nice in order to build up their practices. A detached, crisp, professional, businesslike approach can be substituted for bedside manners. The fact that patients may perceive this approach as cold and depersonalizing has less economic consequences for the specialist than such an interpersonal style would have for the general practitioner.

The Physician as Employee

A final factor that reduces the physician's dependence on patients is the growing trend for physicians to be salaried employees rather than self-employed. Today, increasing numbers of physicians are employed by hospitals, medical centers, Health Maintenance Organizations (HMOs), and other corporate entities that provide public health care. For the physician, there are many advantages to being an employee. Generally, salaries are high. The physician does not have to pay the start-up costs of a new office, furnishings, and medical equipment—nor hire a nurse and receptionist. Finally, the physician does not have to go through the frequently laborious process of building up a patient list that will provide a decent income. Patients are provided by the employer, who advertises, staffs offices, and covers all overhead costs in exchange for the physician's services.

The emergence of salaried physicians in large numbers further weakens the ability of patients to exercise influence because the physician is directly dependent on his or her employer, rather than the patient, for income. Thus, the simple equation for describing the balance of power between physician and patient (i.e., the fee for services) is weighted against the patient. The physician is economically dependent on the employer, not the patient.

Such employment weakens the patient's power in other ways, as well. For instance, an efficient use of physicians is to assign patients to physicians on a first come, first served basis. It is far less efficient to match particular patients with particular physicians. As a result of this practice, patients in hospitals, HMOs, public health centers, and corporate group practices may see a different physician on each visit. For the physician, this policy creates a factory-like atmosphere of processing caseloads. Too much time spent with one patient may be seen as inefficient by the physician's superiors.

For the patient, this organizational policy creates a sense of powerlessness. He or she has to explain to the new physician once again the problems. Questions about trust and about whether the new physician understands or has the ability to treat their illnesses serve to reduce patients' confidence in the process of treatment and in the influence attempts of the physician.

Consequences for Patient–Doctor Relations

Modern medical technology provides physicians with both a base of power and strong tactics to influence patient outcomes. In this climate where the balance of power now favors the physician and technology provides strong ways of controlling disease, it is to be expected that affection between physicians and their patients should also change for the worse. Shorter comments on this worsening climate have been written as follows[2]:

In the 1980s more than a billion encounters between doctors and patients occurred in the United States. On an average, each person had five contacts with a physician during the year. *Many of these contacts ended with anger and frustration on both sides.* (p. 295)

Summarizing many surveys among patients of their feeling about treatment, Glin Bennet,[7] a physician himself, also reports that

there are deep [patient] dissatisfactions at the human level. This is borne out by numerous surveys of the quality of communication between doctors and patients and the feelings patients voice about the quality of care they receive: overall a good two-thirds of patients express dissatisfaction in this area. (p. 68)

Why the anger and frustration? In what ways has the physician's added control fueled this lack of affection? From the perspective of social power, I suggest that the anger and rage result from the changing perceptions that physicians now hold of themselves and their patients. Patients are less often seen as individuals of equal worth. At a minimum, I would expect that physicians would prefer to restrict the amount of time they spend with their patients.

This kind of preference by the more powerful has been found in many other settings. For example, University of Michigan psychologist A. Cohen,[8] reported that, when given a choice, more powerful persons in work settings prefer to talk with individuals of equal power. If required to talk with less powerful persons, they speak for shorter periods of time, reveal less about themselves that is intimate or personal in nature, and try to limit their conversations to work-related issues. No personal chitchat of any consequence can be noted. To make matters worse, other research has found that less powerful persons want to prolong conversations with the more powerful and to exchange intimacies. Both parties, then, must find conversations frustrating. In short, we move away socially and psychologically from the less powerful. Extending Cohen's conclusions to medical relations suggests that today's physicians would prefer to restrict social time with the patient.

To personalize Cohen's account, I recently had surgery of a moderately serious kind that involved staying in a university hospital for 6 days. Each day, the surgeon, nurses, and lesser physicians in various stages of training swept by my bed on their daily visits. Neither they nor I had much to say to each other. The surgeon examined diagnostic signs from charts and from machines connected to my body. He spoke knowingly to his following; occasionally he used my body to illustrate a point. To give him his due, he always asked me how I was feeling. I would mumble, "Fine." I felt strangely concerned to keep his goodwill. If something had gone wrong, I would have had the impulse to apologize for ruining his statistics. I have no idea how he felt about me. I suspect he was as indifferent to me as I am to a grocery store clerk. And yet, one would think that some kind of intimacy should exist between us, because he had seen me naked and thrust a knife into me.

Support for the conjecture that imbalances in power have changed communication between physicians and patients is found in a recent survey of 430 primary care physicians in Massachusetts. Forty-two percent were unwilling to counsel patients about smoking, 54% unwilling to counsel about alcoholism, and 71% unwilling to counsel about life stresses.[2]

Many modern specialists prefer simple diagnostic questionnaires that allow patients to answer "yes" and "no," rather than to waste time interviewing in depth. The chairperson of the American Society of Internal Medicine's Practice Management Committee argued that patients should be persuaded not to insist on talking with the doctors themselves. A paramedic could just as well take a patient's life history. If not a paramedic, then laboratory tests can provide far more information about the patient's health than can speaking to the patient.[2]

Thus, avoiding giving advice, and using questionnaires, paramedicals, and laboratory tests represent ways in which physicians can avoid prolonged social contacts with patients.

Another means of avoiding intimacy, if personal contact is required, is for the physician to use technical language. We have already mentioned that there is a long tradition in medicine of using language to assert the physician's status and power. This tradition continues today. That is, physicians today talk about blood chemistry, renal functioning, and the like. Needless to say, such conversations are beyond the intellectual understanding of most patients. This lack of understanding may further convince the physician that he or she is an effective and insightful person whose knowledge is superior to that of others. At a minimum, the use of technical language serves directly to keep patients at arm's length.

In a study of patient–doctor communications, for instance, Korsch and Negrete[9] found in Los Angeles Children's Hospital that 20% of the mothers did not understand what was wrong with their children, 50% did not understand the origins of their children's diseases, and 42% did not follow the advice they received because they could not understand it.

Needless to say, the many ways in which physicians attempt to limit conversation are experienced as rebuffs by most patients. The anxiety, fear, and uncertainty they experience goads them to obtain information from their physicians, and yet the same information seeking is perceived by the physician as unwarranted and unnecessary. The consequent anger and rage of both parties is understandable.

Ivan Illich[5] describes the impasse in communications between physicians and patients as follows. The patient is

no longer an individual, but a set of symptoms. His condition is interpreted according to a set of rules, and in a language that the patient is less and less likely to understand. Freed by technology from having to concern himself with the whole person, the doctor now can focus on specific disease sites. His recommendations can be based on information provided by machines and laboratories tests, rather than information provided by the patient. (p. 136)

Control is also reflected in physicians' beliefs in the infallibility of their own judgments. At times, this certainty of belief results in physicians using medical technology to treat patients without the patients' prior informed consent. Glin Bennet[7] offers the following vignette to illustrate how power has decreased physicians' concerns for patients' feelings:

A married women of forty-eight was admitted for treatment of breast cancer. The plan, as she understood it, was that she would have a biopsy under general anaesthetic, but when she woke her left breast was gone. She was told the growth was more extensive than had been anticipated. Next day she complained to the surgeon that she should have had a change to think things over once the biopsy report was available .... The surgeon merely said, "What's a breast, anyway? It's only icing on the cake." (p. XVI)

I do not mean to imply that many or all physicians deliberately become callous or indifferent. Most patients express satisfaction with the technical care they receive. Furthermore, almost all medical schools include courses designed to alert medical students to the needs and psychology of patient care. My point is that the common human failures of physicians are the inevitable result of the fact that power changes social relations between the more and the less powerful. What has happened, in this instance, is that the dependency of patients on doctors has increased. At the same time the dependency of the physician on the patient has been reduced. As a result, even with the best of intentions, physicians' attitudes and evaluations of patients follow the historically charted path of social distancing and depersonalization of relations.

Future Trends in Medicine

I started this chapter by asking why family physicians of the early decades of this century were highly regarded by most Americans. The answer I offered is that the physician's public personality was created by the medical technology of that time. Unable to dictate terms to their patients, having uncertain control over disease, and less control over competition, physicians had to present themselves in such a way as to develop dependency, trust, and loyalty in their patients. Out of these needs developed the physician's compassionate style.

We have seen that the physicians' style of presenting themselves has changed from that of a family counselor with compassionate bedside manners to that of an impersonal professional who moves from patient to patient with an entourage of interns, diagnosing and prescribing within short moments of patient contact. Today, few medical practitioners in their right minds would dream of making home visits, spending long hours building trust and confidence, and continually calling and checking to see that the patient followed the prescribed medical regime. There is no need for all that. Modern medical techniques have increased patient dependency by many times and allow the practice of medicine with the precision and impersonality undreamed of in the 1900s.

Rephrasing Marx's views of the links between technology and social relations, it may be said that as inoculation and ideas about hygiene gave us the kindly physician of the 1900s to 1940s, antibiotics, new diagnostic equipment, and innovative operative techniques have created the modern impersonal doctor. Today's physician can cause real improvements in patients' health. Using the discoveries of biochemistry, physics, and genetics as their bases of power, many are rulers of small kingdoms of nurses, laboratory technicians, support staff, and young interns. The patient's role is that of supplicant. The status of medicine has truly risen from its early days.

But what of the future? Technology today has left the physician in charge. Will it tomorrow? As I mentioned earlier in this chapter, physicians in increasing numbers are salaried employees of hospitals, residential care centers for the elderly, HMOs, and chains of neighborhood doctors' offices. It remains to be seen whether physicians as employees can successfully maintain their unique control. Technology already can duplicate many of the diagnostic and healing skills of physicians. These newer techniques provide the potential for fractionating the physician's work along traditional industrial lines, thus providing a significant dollar savings for organizations that employ physicians. Thus, management may replace physicians with technicians whose job will be to manage diagnostic equipment and the computer's diagnosis and treatment plans for patients. If these events happen, the next decade may find a two-tiered system of health care specialists. The first tier will be composed of technicians, who will diagnose, oversee, and treat sickness of a routine kind. The second tier will be composed of the few medical specialists needed to perform the labor-intensive work of medicine involving operative procedures and unusual medical events. In this scenario, technology, once the physician's ally in gaining power, now threatens.

References

1. Starr, P. (1982). *The social transformation of American medicine.* New York: Basic Books.
2. Shorter, E. (1985). *Bedside manners: The troubled history of doctors and patients.* New York: Simon and Schuster.
3. Edwards, M.H. (1972). *Hazardous to your health.* New Rochelle, NY: Arlington House.
4. Reiser, S. (1978). *Medicine and the reign of technology.* Cambridge: Cambridge University Press.
5. Illich, I. (1976). *Medical nemesis.* New York: Random House.
6. Ehrenreich, B., & English, B. (1978). *For her own good.* (pp. 74–78). New York: Doubleday.
7. Bennet, G. (1987). *The wound and the doctor: Healing technology and power in modern medicine.* London: Secker and Warburg.
8. Cohen, A. (1958). Upward communications in experimentally created hierarchies. *Human Relations, 11,* 41–53.
9. Korsch, B.M., & Negrete, V.F. (1972). Doctor-patient communication. *Scientific American, 227,* 66–75.

CHAPTER 6

The Routinization of Work

Many critics of modern production processes assume that as technology reduces the ability of people to demonstrate competence, there are concurrent losses in peoples' feelings of self-esteem and personal worth. From Karl Marx in the 19th century to Harry Braverman today, a core argument against the mindless application of technology to the process of work focuses on these concerns.

The empirical evidence in support of this view is, I believe, mixed. Robert Blauner,[1] for example, in his classic study of alienation among factory workers in chemical, printing, textile, and automobile industries, found that social alienation was highest in the latter two industries, in which workers had little freedom or autonomy. Yet at the same time, Blauner found that workers in an automated chemical plant showed a far lower degree of subjective alienation than those in the textile factory or an automobile production line. Paradoxically, the increased automation in the chemical plant may have resulted in work being so organized that employees could exercise greater control and skill over the production process. More recently, Shoshana Zuboff[2] in her studies of the introduction of computer technology into the manufacturing process, also reports instances in which the new technology increased rather than decreased the employee's ability to exercise control. Similar findings are reported by Barbara Gutek and her colleagues in their analysis of the introduction of computer-based office information.[3] The majority of office workers reported more rather than less satisfaction with using the new computer equipment. To further confuse any simple conclusions, studies by two organizational psychologists, Greg Oldham and Richard Hackman,[4] report that routinization of work is only troublesome for employees with strong needs for self-actualization and self-development. For the remaining workers studied, Oldham and Hackman report that routinized work did not affect employee job satisfaction for the better or the worse.

In short, there appears to be no simple direct linkage between technological changes in work and the responses of workers. The large-scale alienation described by Marx is difficult to detect by empirical study. Workers' responses to technology appears to be mediated by many events, including both the extent to which the new technology increases or decreases worker control over the process of work and individual differences in the need of workers to satisfy psychological growth needs, as described by Oldham and Hackman.

Technology Reduces Management's Dependency on Labor

What I believe may be easier to detect are changes in managers' social relations with their employees as a consequence of technology's reducing employees' control of the work they do. When this happens, managers are less dependent on employee skills and talents in order to get the work done. That is, managers, rather than their employees, control the production process.

A description of the introduction of computers into office work by two social psychologists, Sara Kiesler and Tom Finholt,[5] illustrates some of ways in which it is possible to use technology to increase managerial control of employee performance. These authors were investigating the possible psychological origins of repetitive strain injury among office workers in Australia.

Computer power allows management to increase productivity using means that also increase monotony, muscle strain and fatigues. For example, with the introduction of computer data entry and work processing, files are accessed through the computer rather than kept in filing cabinets. This makes it possible to cut the number of work breaks and increase the pace of work, and managers can monitor these activities automatically using the computer. When the Australian Taxation Office introduced computers in the 1970s, it progressively increased productivity by increasing the rate of data entry; by 1981 data processing clerks were required to make 14,000 keystrokes per hour. (p. 1007)

What can we expect as a result of these changes in management's ability to control? The metamorphic model suggests two changes in managers' relations with their employees that I will examine in the following section. First, managers should give employees little credit for satisfactory work. Second, managers should blame employees for unsatisfactory work regardless of the actual reasons for the poor work.

Credit for Good Work

There are several related reasons why routinization may alter managers' evaluations of employees who do satisfactory work. As illustrated in the previous paragraphs, the pace of employee work can be regulated by machines. Furthermore, employee performance can be immediately evaluated, work can be monitored, employees' discretionary powers to plan and make decisions are slight, and minimal cognitive skills are needed to do the work. Under these circumstances, there should be little doubt in the minds of managers that employees are not in control of their work-related behavior. It is far more reasonable to attribute satisfactory work to the effects of such external causes as machinery, the system of work, and the manager's direct orders.

Let us suppose, for example, that we are in a modern machine shop. Here, we watch the operations of a computer-controlled lathe. First, we see a worker place an unfinished piece of metal in the stocks and press a button. Then the lathe makes a series of complex cuts into the metal. Soon the lathe stops its operations and the worker removes the finished metal from the stock. Surely, we would not compliment the worker on his skill in fashioning the metal into the finished product. We all know that the computer-controlled lathe was responsible for this good work.

Blame for Poor Work

While we may not praise employees for satisfactory work in automated settings, we surely will blame them for poor work. Suppose, for instance, a new automated cash register is installed in a fast-food restaurant. The cashier has simply to hit a picture of the particular hamburger chosen by the customer, read off the amount owed by the customer, record the actual money received from the customer, and finally, hand back the correct change as indicated by the cash register — not quite a totally automatic operation, but close. Certainly, the cognitive elements of the job involving adding and subtracting have been eliminated. Let us further suppose that the cashier, using this new system, continually returns too much money in change to customers.

Now the manager, on becoming aware of the fact that the cashier is returning too much money, has two sources that can be blamed. The first is that the computer register is broken; that is, it is systematically making errors in computing change. This possibility can readily be checked. The second source of blame is the employee. If the employee is blamed, then the manager must decide whether the work itself is too difficult and the cashier needs additional training, or whether the errors are being made deliberately.

In this instance, the manager might find it difficult to attribute the errors to the fact that the work is too hard or complex. No arithmetic is required. The work has been so simplified that any one is able to do the work. Thus, the manager can only conclude that the cashier's mistakes have been deliberately made. Among the probable reasons are the cashier's poor attitude, lack of motivation to work hard, or even deliberate attempts to sabotage the system.

Technology, then, as it routinizes work, can be described as enmeshing individuals in a system in which they do not appear to be in control of their own behavior; and in which individuals are not given credit for acceptable performance, yet run the risk of being assigned responsibility for poor performances.

Field Studies in Industry

For the reminder of this chapter, I shall describe field research in offices and factories, and a laboratory simulation of the work setting, that have attempted to verify whether managers' attitudes toward, and social relations with, their employees are consistent with the above statements.

In these studies I wanted to test the following ideas about social relations between managers and their employees as technology routinizes work. The following is the kind of information I was looking for and what I expected to find:

1. Managers of routinized units would evaluate their employees' performances unfavorably as compared to managers in nonroutinized work units (Studies 1,

2, 3, and 4). My assumption here was that managers of routinized work units would credit technology rather than employee effort for any satisfactory work that is done. Hence, managers would denigrate the contributions of their employees and evaluate their performances unfavorably.

2. Managers of routinized units would believe that unsatisfactory employee performance was caused by their employees' poor attitudes and lack of motivation. On the contrary, managers of nonroutinized work would believe that unsatisfactory employee performance was due to the difficulty of the work (Studies 3, and 4). As I mentioned earlier, if the employee's work has been simplified so that the machine does all the difficult work, then the most logical explanation for poor work is employee malingering and malice.

3. Managers of routinized units would use strong verbal tactics more frequently and rational tactics less frequently when giving orders to their employees (Study 3). Here I take a closer look at the kinds of authority relations that develop out of routinization. In routinized settings, I assume that employees have little power. Hence, there are few costs for the manager if he or she yells and demands compliance (i.e., uses strong tactics). Furthermore, since the work is not difficult, managers do not have to take time to explain what they want their employees to do. Basically there is nothing to explain.

4. Managers of routinized units would use strong verbal tactics less frequently to influence their superiors than managers of nonroutinized work units (Study 3). Organizational psychologists Gerald Salancik and Jeffrey Pfeiffer have continually found that managerial power is contingent on the management of uncertainty.[6] Those managers directing work activities for which there are no ready solutions have more power than those directing units that present few problems. Thus, managers of a secretarial work pool have less organizational power, for example, than managers of a group of design engineers.

In short, not only may workers lose power in settings where technology has routinized their work, but the managers of such units will also lose power to influence their own superiors. This is because their unit's work is predictable and presents few production problems. In comparison, managers who supervise work that is complex (i.e., in which uncertainty is high) become relatively influential and powerful because other units depend on their expertise to solve important problems. Lacking organizational clout, then, managers of routinized work units would want to "walk softly" and avoid confrontations with their own bosses. In practice, this means avoiding the use of strong and demanding influence tactics when trying to influence their superiors.

Some Empirical Findings

Our first two studies gathered information from small groups of first-line supervisors. These supervisors described the extent to which the employees in their units did routinized or nonroutinized work. They next evaluated their employees' job performances. There were 34 first-line supervisors in the first study and 22 in the second study.

The first field study used information collected by Charles Perrow in the late 1960s as part of a large study of technology in business firms.[7] One segment of Perrow's study examined the attitudes of production supervisors. I subsequently reanalyzed this segment.

Perrow measured the extent to which work was routinized by asking each supervisor two questions: (a) whether the work done by his employees produced predictable results and (b) whether his employees could routinely solve day-to-day problems as they arose. Supervisors who said that their employees' work was predictable and that day-to-day problems could be easily solved were classified as supervising employees who did routine work. Perrow also asked the supervisors to evaluate their subordinates in terms of the amount of (a) effort they showed, (b) pride in their work, and (c) loyalty to the company.

The second study was based upon the answers of 22 first-line supervisors in various Australian business organizations. In this survey, the number of questions about the routinization of the work was increased to four: (a) was the work standardized, (b) could work problems be easily solved, (c) were repetitive work procedures used, and (d) could the production output be easily forecast. Similarly, the number of questions evaluating employees was increased to five. We asked the supervisors to evaluate their employees in terms of: (a) employees' attitudes toward the work, (b) the need to continually supervise each employee, (c) the quality of the work done, (d) the quantity of the work done, and (c) the amount of employees' pride in their work.

The findings of these two studies supported the metamorphic model. In both studies we found that as the work done by the work unit increased in routinization, the supervisors' evaluations of their employees became less favorable. In the first study, the correlation between the measure of routinization and supervisors evaluations of their employees was $-.34$. In the second study, it was $-.41$.

In everyday terms what this means is that as work is simplified and its skill content removed supervisors are increasingly likely to express negative feelings about their employees. That is, supervisors of employees doing routinized work supported such opinions as that their employees "had to be watched closely," "took little pride in their work," and "had poor attitudes towards the company." On the contrary, supervisors of employees doing complex work praised their employees' motivations, attitudes, and pride in their work.

Increasing the Number of Supervisors Surveyed

So far, we have some evidence to support the assertion that when technology takes control away from people they are evaluated less favorably than those who remain in charge of their work-related behaviors. One should keep in mind that these were not evaluations of the actual work itself, but of the employees' motivations and attitudes toward work. These aspects of the work should not vary with the job's complexity. In theory, we can have sullen craftsmen and well-motivated assembly line workers.

The critical reader might, however, wish that the number of supervisors studied could be increased. In addition, one could ask that we look in more detail at the ways in which supervisors' opinions of their employees changed as the complexity of their employees' work changed.

For these reasons, we conducted a third survey among 75 first-line supervisors employed in a spectrum of occupations in the Delaware Valley of Pennsylvania. Their occupations included the supervision of direct production workers, maintenance, shipping, sales, personnel, and finance. The majority (69%) were in direct production.

To measure the extent to which the supervisor's employees did routine work, we asked much the same questions as in the second study. Only this time we also included questions about the need of employees to have apprenticeship or trade school training, the amount of formal education employees needed, and how long it might take employees to learn to do their jobs well. These questions assumed that the more training needed and the more time it took to learn the work, the less routine the work.

We again asked supervisors how favorably they evaluated their employees. This was done by asking the supervisor to describe the extent to which their subordinates (a) took pride in doing good work, (b) needed to be closely supervised, and (c) accepted responsibility for their work.

As in the first two studies, we found that the more routinized the work done by employees, the less favorably the supervisor evaluated them. In this instance the correlation between the extent of routinization and supervisors' evaluations was $-.40$. The more routinized the work, the more likely supervisors were to describe their employees as taking no pride in their work, as needing to be supervised closely, and as not being responsible for the work they did.

Thus, in three separate surveys we find a consistent hardening of evaluations toward employees by management as the work becomes simplified. Such findings are consistent with the idea from the metamorphic model that we perceive and evaluate unfavorably people who are not in control of their own behavior.

Explanations for Poor Work

It can be recalled from the earlier discussion about a cashier who gave out the wrong change that there are several alternate explanations of the causes of poor work that an outside observer can make, if one rules out breakdowns in the machinery. The first explanation is that employees deliberately ruined the work because of hostility or perhaps a "just don't care" attitude. We expected these explanations to be used more frequently by supervisors of employees doing routinized work. This is because it is difficult to talk about lack of skill as a cause of poor work when no skill is involved in the work to begin with.

The second and more charitable explanation is that the poor work was due to the difficulty of the work or the employee's lack of ability. In this second explanation, poor work is seen as accidental and not due to the employee's deliberate fault. Given that a fair measure of skill is required to do nonroutine work, we expected supervisors of such units to give their employees the benefit of the doubt and attribute poor work to problems of skill.

To measure these different explanations, we asked the following two questions of our 75 supervisors:

(a) About what percent of your subordinates, if any, have a basic weakness in their work because they lack the ability to do the work? (Answers ranged from 0% to 30% and more.)
(b) About what percent of your subordinates, if any, have a basic weakness in their work because of poor attitudes and/or a lack of motivation? (Answers again ranged from 0% to 30% and more.)

What we found was that when the work that employees did was complex—that is, when the work was challenging, nonroutine, involved varied components, and required planning—supervisors described employees who were weak in their work as lacking in ability. The correlation between the measure of routinization of work and estimates by the supervisor of the number of employees who did poor work because they lacked skill was $-.40$.

Fewer supervisors attributed poor work to this reason as the work became routinized. The latter supervisors had a different explanation for the poor work done by their employees: when the work that employees did was routinized and paced by machines and machine systems, employee weaknesses were attributed by supervisors to the employees' lack of motivation and poor attitudes. The corresponding correlation between the measure of routinization work and the explanation that this poor work was due to poor attitudes was $+.22$.

In short, when poor work occurs in routinized work settings, managers believe that it is due to the lack of motivation of their employees. When poor occurs in nonroutinized work, managers believe that it is due to employees' lack of ability, certainly a less harsh judgment.

A question of some interest is whether these findings truly describe the behavior of employees. Are weaknesses in work really caused by employee hostility in routine settings? Perhaps yes. It may be that the most poorly socialized members of the working class are the only kind we can get to do routine work. If this is so then the supervisors' judgments are accurate. Poor work is due to their employees' hostile attitudes. Perhaps, however, it is not true. Perhaps these attitudes are only in the supervisor's head. By this I mean that when work is routinized, as I suggested earlier, it fosters the illusion that only deliberate mistakes can be made in the work.

Perhaps it does not matter which explanation is correct. In fact, probably both are correct. That is, as management routinizes work it is in a position to employ the least expensive segment of the working force, and this segment probably contains more frustrated and hostile people than the segment of skilled workers. But it is also true that routinization produces the perception that employees cannot make mistakes. The system is "idiot-proof," goes the saying. Consequently, if mistakes do occur, the outside observer's most reasonable explanation is that malicious intent was involved.

I shall return to this matter at the end of this chapter. Suffice it to say here that the grounds for deteriorating social relations between people at work are clearly in place when we assume, truly or not, that one's employees have deliberately set out to do the organization harm.

Managerial Power and Routinization

Earlier in this chapter, I pointed out that managers of routine technology also lose the ability to influence others in the organization. This is because the work they supervise is predictable and, hence, not a critical problem area for the organization. Thus, their work is taken for granted.

To examine whether routinization caused supervisors to lose power in the organization, we asked each of the 75 supervisors in our survey about their ability to exercise influence. Each supervisor described the amount of influence he or she had in (a) setting budgets for their work units, (b) coordinating with other units, (c) influencing organizational policy in areas that were important for their work, and (d) influencing higher-ups in the organization.

We found that supervisors who said they had very little organizational clout were, in large proportion, those who described their units as doing routine work. In contrast, supervisors directing units doing complex work felt they had a fair amount of say in controlling organizational policies that directly affected their units.

Perhaps this is not a surprising finding to people familiar with organizational theory and politics. As I have mentioned, organization theorists, such as Salancik and Pfeiffer,[6] have pointed out for some time that power in an organization is contingent on the management of uncertainty. That is, in the same organization, the supervisor of a group of clerk-typists has less influence than the supervisor of a group of design engineers. This is because the work of clerk-typists presents fewer problems of importance to the organization than does the work of engineers. Pursuing this matter a bit more, I suspect one reason why organizational power shifted from production to finance, in such major industries as the automobile industry, in the 1960s (see, for example[8]) was that manufacturing problems became routinized during this period. As a result, organizational control shifted to managers involved in the increasingly uncertain problems of finance caused by international competition, mergers, and industrial expansion.

How Orders Are Given in Routinized Settings

If people are powerless and we want them to do something, the metamorphic model suggests that we don't waste time explaining what we want. We simply tell people what to do. By the same token, people who are powerless should confront authority in a passive and meek way, so as not to give offense.

In chapter 2, I described how we measured the various dimensions of verbal influence used by managers in industry to influence either their subordinates or their superiors. These dimensions include the use of strong tactics (I demand), weak tactics (I act humble), reason (I explain my reasons), bargaining (If you do this for me, I'll do what you want), coalitional pressures (I bring in my

TABLE 6.1. Correlation between influence strategies used
by supervisors and routinization of the work unit.

	Strategies used to influence employees	Strategies used to influence superiors
Weak tactics	−.06	.05
Reason	−.24[a]	−.29[a]
Strong tactics	.12	−.25[a]
Sanctions	.10	[b]
Bargaining tactics	−.06	−.18
Higher authority	−.02	−.45[a]
Coalitions	−.26[a]	−.13

[a] Indicates the correlation is statistically significant.
[b] Not measured for superiors.

coworkers) and appeals to higher authority (I speak to my boss's superior). An
additional dimension, the use of sanctions (I threaten poor performance evaluations), was found to be used when influencing subordinates, but not superiors.

The 75 supervisors from the Delaware Valley were asked how frequently they
used each of these tactics when influencing, first, their subordinates and, then,
their own bosses.

What we expected to find was that the supervisors of employees doing complex
work would rely on logic and reason when influencing both their subordinates
and their bosses. On the contrary, we expected supervisors of employees doing
routinized work to use strong tactics and sanctions when influencing their subordinates, but to be nonassertive (i.e., use little influence) when trying to influence
bosses. These different patterns of using influence should emerge if technology
does alter the balance of supervisory power in ways that I have suggested.

What we found is summarized in Table 6.1. Here I show the correlation
between each supervisor's use of the seven influence strategies and the extent to
which the work of his or her unit was routinized.

The reader may immediately note the many negative signs in front of the
correlation coefficients. This means that supervisors of routinized work units
were *less* likely to try to influence either their subordinates or superiors than
were supervisors of nonroutine work units. For example, the correlation of −.24
means that supervisors of routinized work units used the tactic of reason less frequently than supervisors of nonroutinized work units when attempting to
influence their employees.

These findings about face-to-face influence tactics support, if not fully, the
belief that technology is implicated in the kinds of social relations that emerge
between management and labor. Originally, I thought that supervisors of routinized work units would use strong and controlling tactics to influence their
subordinates. That is, they would demand and threaten. Such tactics could be
used safely because supervisors would not be concerned about giving offense to
their unskilled employees.

We did not find this to be true, although the sign of the correlations for strong tactics and the use of sanctions were in the predicted direction (+.12, +.10). Instead, supervisors of routinized work units were relatively passive in their attempts to influence both their subordinates and their superiors. That is, they avoided the use of reason and the use of coalitional pressures with subordinates, and avoided the use of reason, strong tactics, and appeals to higher authority with their own bosses. The best explanation I have for their passiveness in exercising influence is that supervisors of routinized work felt powerless to intervene, either with their employees or with their bosses. In this instance, then, the routinization of work may have left both employees and those who supervised them relatively powerless.

Questions about Cause and Effect: A Laboratory Simulation

Social scientists are quick to tell us that survey studies do not prove anything about the causes of behavior. Rather, such studies describe what is currently happening—valuable information in itself, but not convincing if one is trying to prove a point, as I am.

I will draw upon an experimental study of routinization to further illustrate, and perhaps support, my arguments. In this study 52 college students were assigned to serve as supervisors of subordinates doing either routinized or non-routinized work.

The students first met in groups of three in a room set up as a production factory. On a table were many containers of different colored Lego blocks. There were also assembled models of the company's product—airplanes and cars. The students were told that one of them would be a manager and the other two students were to be production workers. Each production worker would be in a separate manufacturing room and the supervisor would receive periodic reports of the performance of each of the two workers.

Students were shown either a routinized or nonroutinized way of manufacturing the model airplane. In the routinized work there was a 4-foot-long conveyor belt. The students were shown how the worker would snap together Lego blocks and place each unit on the belt. When enough of each unit had been assembled new combinations of blocks would be used until the model was finally assembled.

In the nonroutinized work, students were shown three different sets of plans of an airplane from a Lego model book. They were told that the employee would choose to build any one of the three models. The plans contained no written instructions. Thus, successful performance depended on the skills of the employee.

Having established two different kinds of working conditions, each student was next taken to another room where he or she was assigned the job of manager, while being told that the other two students would work as employees doing the assembly work. The manager was kept busy filling out reports on his or her unit's progress, or reading reports on company policy. Every 5 min each manager received a report on the progress of the work – that is, how many Lego blocks had been assembled and how many mistakes were made. From this information they could calculate the dollar value of each employee's performance.

Since the reports on the work done were simulated, all managers received the same reports about production. In addition, the periodic feedback showed that the work done by the first employee consistently exceeded company standards, while the performance of the second employee was consistently below company standards. In other words, in both the routinized work setting and the nonroutinized work setting, one employee's performance was superior and the second employee's performance was unsatisfactory.

How did the routinization of work influence the evaluations of the student managers about their two employees? Despite the fact that the actual amount of work done by each employee doing routinized or nonroutinized work was exactly the same, our reasoning suggests that the good work of employees doing routinized work might be attributed to the conveyor belt system. Hence the employees themselves should not receive high grades for the work they did.

We found that managers' evaluations of their two employees at the end of an hour of work were quite similar to those of managers in the field surveys. Good work was attributed by the student managers to the talent and skills of the effective employee in the nonroutine work setting, but not in the routine setting. That is, the effective employee in the nonroutine unit was evaluated as far more competent than the effective employee doing routine work despite the fact that effective workers in both settings were doing equally good work in terms of actual production.

How did managers evaluate their ineffective employees? Here the differences in evaluations focused on employee motivations. Both ineffective employees were rated as equally incompetent with regard to their skills and abilities. However, the ineffective worker doing routine work was described as lazier and far less motivated than the ineffective worker doing complex work.

This last set of findings was more fully explained when the managers were asked why they thought their employees did poor work. We asked the following question about each of the two employees: If this employee's work was not satis-

TABLE 6.2. Managers' explanations for their employees' poor performance.

	For employee doing routine work	For employee doing complex work
Employee poorly motivated	77%	27%
Employee lacked ability, work too difficult	28%	73%

factory, would you say that the poor work was due to (a) poor attitudes toward the work, (b) a lack of ability, or (c) the difficulty of the work. Alternatives (b) and (c) were scored as attributing poor work to a lack of employee ability. Alternative (a) was scored as attributing poor work to a lack of employee motivation. None of the mangers said that the performance of the first employee (who did above average work) was "not satisfactory."

The proportions of managers attributing their employees' ineffective performance to either a lack of motivation or to a lack of skill is shown in Table 6.2. It can be seen, as in the field studies, that as the work became routinized, managers were more likely to believe that poor work was due to voluntary reasons, such as laziness, than to involuntary reasons, such as the work becoming too difficult.

Conclusion

The information that has been presented in this chapter supports the idea that variations in technology are associated with changes in managerial appraisals of labor. These changes in appraisal are consistent with predictions made by the metamorphic model. With routinization, control of the skilled components of work are transferred from employees to machines and machine systems. As a consequence, technology, rather than the employees, is given credit for acceptable work. Furthermore, routinization produced the belief in managers that poor performance was due to deliberate malice on the part of their employees rather than to involuntary causes such as a lack of ability or the work becoming too difficult. These kinds of attributions must surely reinforce attitudes of hostility and devaluation among managers.

The laboratory study found the same kinds of social relations emerging from technological control. A further contribution of the latter research was to distinguish between the performance of effective and ineffective employees. What we found was that satisfactory performance is not attributed to employee skills and talents in routinized work units. Why should it be? The skill has been taken from their work. The machine now controls quality—not the employee. At least this is what the manager believes is happening. Thus, our findings may have simply documented the obvious. Yet even the obvious has to be made explicit from time to time, or it becomes an unnoticed part of the background.

I believe that within the limits of information that can be generated by the methods of social science, this chapter has shown that the attitudes and evaluations of managers are clearly influenced by technological forces. There seems no other reason but technology to explain why supervisors did not give their employees doing routine work credit for good work, and why they blamed them for poor work. Holding such beliefs inevitably must cause social relations between labor and management to sour. However, the cause of this increasing social stratification between labor and management may not, as Marx and others have suggested, be due to labor's alienation, but rather to an inevitable hardening of managerial attitudes.

Technology and Democracy

Not surprisingly, technology has implications for the process of democracy. The Jeffersonian ideal is based on a political state composed of independent, critically thinking individuals who exercise rational control over the state's functions. A core assumption of democracy is that people are capable of exercising control over their own lives. Once this assumption is questioned by political leaders, there is a concurrent reluctance to share power.

By democratic process, I do not mean whether or not people vote in elections. Rather, I mean the day-to-day democratic process in which people are consulted and their opinions are taken into account.

The social science literature has increasingly documented in industrial studies that as the work becomes routinized, the working environment becomes authoritarian. That is, employees doing routinized work are not consulted by their superiors, nor are they allowed to decide for themselves how the work should be done. Logically, this makes sense because as workers lose control of their workaday world, they are seen as having little to contribute and, hence, are consulted less. Most recently sociologists B.C. Reiman and G. Inzerilli[9] concluded that

a careful look the results of all the studies reveals a basic convergence about the relationship between work group structure and technology. That is, as work becomes less routine, or more uncertain and complex, its structure becomes more organic, with increased autonomy, participation, and informality of relationships in the group. The results then consistently point to a *clear technological imperative* at the level of the work group. (p. 17)

A similar conclusion is offered by Dutch psychologist Robert A. Roe[10] in a survey of the results of the introduction of computer technology in European countries. While the introduction of computer technology had no systematic effects on such aspects of organizational functions as changes in the numbers of employees supervised by each manager (i.e., span of control), there was a trend among the many European companies studied for control of employees to be concentrated at higher levels of the organization. At the same time, impersonal, mechanical means of control were introduced at the lower levels, "demanding strict obedience to work procedures and rules" (p. 6).

In another paper illustrating this important link between democracy and technology, organizational psychologists L.R. James, M.J. Gent, J.S. Hater, and K. Coray[11] interviewed employees who were doing either skilled or unskilled work. Employees engaged in routinized production work said that they had little influence over the way they did their work, or over their supervisors' decisions, as compared to employees engaged in work that had not been routinized. Their supervisors also independently reported that they consulted less with employees engaged in routinized than in nonroutinized work. In short, democracy flourishes when people are autonomous.

Evidence about the link between technology and democracy is also found in the experimental simulation of technology that was described earlier in this chapter. As part of the employee evaluation form given to each manager at the end of the first hour, managers were asked the extent they would "let the employees decide how to do their own work," in order to run a profitable business.

Managers of the employees doing complex work stated that they would allow employees to decide how to do their own work. Managers of employees doing routine work emphatically disagreed by a 2-to-1 margin. Needless to say, a greater readiness to share decision making with employees by managers is a necessary precondition for the establishment of democratic controls in the work setting. In theory, most people favor this form of governance at work. What gives most managers pause, and certainly was the reason for the answers of the managers in our study, is an unwillingness to share governance with people who are not in control of their own behavior.

The metamorphic model of power helps us to understand the psychological basis for this relation between routinized technology and the use of hierarchical, rather than democratic, controls. That is, as technology shapes managers' perceptions they develop strong resistance to sharing decision-making powers with employees. Where work is routinized, employees are seen as not in control of their own work behavior, not motivated, not involved in their work, and lacking in the expertise to make intelligent decisions. Such employees, I would argue, would be perceived as incapable of participating responsibly in democratic decision making. Far better to simply tell them what to do. The technological imperative described by Reiman and Inzerilli,[9] has a psychological basis.

Furthermore, even if managers were required by national policy (as is true in several socialist countries), top management decisions, or union agreement to enter into participative arrangements with employees, I believe that the presence of routinized technology would cause such democratic arrangements to fail. This is because the process of joint discussion and shared decision making would in fact remain one-sided, favoring managers.

Recent studies of group decision making by Dutch psychologist Mauk Mulder[12,13] illustrates how democracy fails when power is unequal. Mulder found that the democratic process of collective decision making broke down when groups vary in such bases of power as each member's status and expertise. Those people in the group who had higher status or were recognized for their knowledge dominated group discussions and decision making. Those with lesser status and expertise silently assented to this domination. Similar conclusions about the control of decision-making power, variations in expertise, and the democratic process have been reached by other social scientists. In short, the routinization of work, and the consequent reduction in the skills required by employees, creates a psychological climate of social distancing and denigration which effectively bars consultation and participation in decision making between employees and managers.

For better or worse, democracy works best when people are actively engaged in what they do and control bases of power that are valued by others. When this happens, peoples' opinions are given weight. Democracy fails as the belief grows among the elite that people are not worth consulting. Further contributing to this failure is the fact that people without power are less likely to participate responsibly in group processes. Rather, they defer to the opinions of the elite, thus further encouraging the development of antidemocratic sentiments.

References

1. Blauner, R. (1964). *Alienation and freedom*. Chicago: The University of Chicago Press.
2. Zuboff, S. (1988). *In the age of the smart machine*. New York: Basic Books.
3. Gutek, B., Bikson, T.K., & Mankin, D. (1984). Individual and organizational consequences of computer-based office information technology. In S. Oskamp (Ed.), *Applied social psychology annual* ©5 (ch. 11). Beverly Hills, CA: Sage.
4. Hackman, R., & Oldham, G. (1980). *Work redesign*. Reading, MA: Addison-Wesley.
5. Kiesler, S., & Finholt, T. (1988). The mystery of RSI. *American Psychologist, 43*, 1004–1015.
6. Salancik, G.R., & Pfeiffer, J. (1977). Who gets power and how they hold it: A strategic contingency model of power. *Organizational Dynamics, Winter*, 22–35.
7. Perrow, C. (1970). Department power and perspectives in industrial firms. In M. Zald (Ed.), *Power in organization*. Nashville: Vanderbilt University Press.
8. Halberstram, D. (1986). *The reckoning*. New York: Morrow.
9. Reiman, B.C. & Inzerilli, G. (1978). *Technology and Organization: A Review and Synthesis of Major Research Findings*. Paper presented at the Conference on the Functioning of Complex Organizations, West Berlin, West Germany.
10. Roe, R.A. (1988). *New technologies and work*. Paper presented at the 24th International Congress of Psychology, Sydney, Australia.
11. James, L.R., Gent, M.J., Hater, J.J., & Coray, K. (1979). Correlates of psychological influence. *Personnel Psychology, 32*, 563–587.
12. Mulder, M. (1971). Power equalization through participation. *Administrative Science Quarterly, March*, 31–38.
13. Mulder, M., & Wilke, H. (1970). Participation and power equalization. *Organizational Behavior and Human Performance, 5*, 430–448.

CHAPTER 7

The Technology of Coercion

Military power—The ability of states to affect the will and behavior of other states by armed coercion or the threat of armed coercion.

(R. Osgood, *Force, Order and Justice.*[1])

I used to own all our curtains and all the materials that covered the chairs. I used to buy printed linens and things in England and bring them out. It was rather nice to see a nice bit of printed linen on your chairs

(British colonist, India[2])

Using technology to increase the certainty of harm-doing is older than civilization. Perhaps the first hominid who discovered how to chip stones into axes deserves the credit for beginning this technology. Each new advance since then has increased the certainty of harm-doing as well as altering the balance of power between contending parties.

At first, the time between the invention of new techniques of war could be measured in thousands of years. The skills needed to fashion bronze into armor occurred perhaps from 4,000 to 6,000 years ago. Yet it took another 2,000 years before the skills that were needed to transform iron into weapons became widespread. The industrial revolution, however, changed this slow pace. The wedding of science, applied skills, and mechanics in the 17th and 18th centuries caused the pace of the development of new techniques of war to accelerate at an exponential rate. Today, planes, missiles, poison gases, germ warfare, nuclear bombs, and more crowd each other for their place in strategic planning. Some would argue that the success of technology in increasing the certainty of killing has overreached itself, in that today we have weapons that we are afraid to use. We are stalemated by our own successes.

In the situations described in the preceding chapters, the power element in technology was indirect. By this, I mean the ability to control individuals through technology was, for the most part, an accidental by-product of attempts to make events more predictable and controllable. In one instance, social scientists and physicians gained power because of the increased dependence of people on their knowledge. In the other instance, managers gained power as technological innovations reduced the skill requirements of employees. But in neither instance was the main goal of technology to provide powerholders with strong means of controlling people.

Weapons, however, represent a technology whose main goal is to control human behavior by making people do what the powerholder wants through threats and intimidation. Using such means leaves no doubt as to who is the master and who is the slave. The robber who points a gun while making demands has little doubt that the gun makes his victims hand over their cash. As long as the robber has the gun, all parties involved would agree that he is in charge.

If we increase the number of guns to regimental size and keep pointing guns at many people for many years, then we no longer have a robbery, but a colony or an administered territory. Surely, we should find in the colonialist all of the many attitudes and behaviors that are associated with the metamorphic model of power. The formula of power in this instance is very simple. Better technology makes for superior weapons. With superior weapons, nations can control people in other lands, exploit their labor, use their resources, and so enrich themselves. Today, as technology has improved weapons, transportation, and communications, Western nations have been able to occupy and control regions of the world many thousands of miles away from their own countries. In this chapter, I will use the words of administrators and settlers living in occupied territories to show how the absolute control provided by weapons leads to almost unlimited power and, in most instances, unrelenting derogation of those who are controlled.

Why Use Coercion?

I begin with the naive question as to why nations willingly spend unlimited money to develop better weapons. Why do we want to deliver noxious stimuli to people who in many instances we do not even know? As with the development of other technologies, the answer has to do with the press of our desires. In this case, however, what we want is controlled by other people. Perhaps we want others to leave us alone, or we may want their possessions, or perhaps we want, quite simply, to harm them. The development of effective weapons increases the certainty that we will achieve all of these outcomes—and they are achieved without wasting time in tedious negotiations. With weapons, we can do what we want regardless of the consent of those we are seeking to influence.

Without technology we might still be following the example of the 18th century primitive Tahitian and New Zealand Maoris warriors, who, to gain courage and victory, were reported by early travelers to eat the hearts of their enemies. Except for its psychological effects, this form of cannibalism has no consequences for victory in warfare. On the other hand, each advance in the technology of weapons building does, in fact, increase the certainty that weapons will achieve their intended effects of harm-doing and intimidation.

I will begin, then, with a brief account of some of the major reasons given in the social science literature to account for coercion. By this account, we can see the many forces acting on decision makers to continue weapons building, despite the pleas of pacifists and others who argue that weapons encourage rather than discourage wars. The multitude of reasons for weapons defies all arguments in favor of disarmament. If one reason for weapons building in considered illogical, still other equally powerful arguments in its favor will emerge.

To Harm Other People

The first reasons for using coercive power emerges out of an individual's rage and anger. The discoveries of centers within the brain that govern the expression of aggression, as well as the sociobiologist's descriptions of evolutionary reasons to harm and to aggress, support the belief that there is a genetic and physiological basis for rage and anger.

The goal of harm-doing, in this instance then, is to force target persons into negatively valued regions in order to humiliate, to cause pain, to torture, and to kill. This perspective answers the question of "Why aggression?" by saying that we enjoy the suffering of others. "Homo homoni lupus," wrote Freud[3] in despair after the World War I, "Man is like a wolf. Who, in the face of all of his experience of life and history will have the courage to dispute this assertion?" (p. 24).

And so we read, hear, and are fascinated about tales of violence in which bloody deeds are done for seemingly no other reason than to inflict pain – "I was jealous," "I hated," "I lost control.") Yet these kinds of explanations can hardly account for the many instances of coercion that are not based on anger and rage. Furthermore, this explanation of aggression is not likely to explain the collective efforts of states to produce better weapons. The enormous developmental and maintenance costs would hardly be justified if weapons were used simply to make the expression of anger more productive. We must look elsewhere to explain why corporations and nations do harm.

To Protect Ourselves

One such reason for coercion that justifies weapon building emerges out of fear and the perception of anger. When fear is aroused, one of the first strategies of most people is to attempt to mobilize greater force than the potential attacker. Few would question the moral rightness of possessing weapons for defense. Former president Ronald Reagan argued, for example, that we must have space-based laser weapons to defend the United States from the first-strike missiles of Russia.

Unfortunately, it is difficult, once armed, to avoid the temptation to use weapons for offensive rather than defensive purposes. Social psychologists consistently report that the possession of the means to harm others, regardless of whether these means were obtained for offense or for defense, leads to aggressive behavior. Once armed, individuals and nations realize that they can shoot first, and so eliminate enemies. Fears tend to exaggerate the target person's potential for harm-doing. This perceptual distortion paves the way, in many instances, for counteraggression. Thus, the successful mobilization of means of defense does not necessarily alleviate fears, but can often intensify them.

To Prove our Worth

Still another reason to do harm is to satisfy ego needs related to a sense of worth. The envious Iago appears to have had no other reason for his destruction of Othello than an inner rage that the Moor was more respected than himself. Thus, to destroy Othello was to reaffirm the worth of Iago. Black psychiatrist Frantz Fanon[4] argued that only the use of coercion and violence would free the colonized from their ingrained feelings of inferiority. In essence, the use of violence becomes a cleansing force, ridding the colonized of their habitual deference and bent knee. One of the many arguments for rearmament in Germany after World War I was based upon the need to erase Germany's humiliation of defeat.

To Take from Others

Still another reason for the use of coercion is based on rational calculations as to how best to obtain a greater share of the good life. Doing harm, protecting themselves, or raising their self-esteem are secondary to the desire of powerholders to enrich themselves by forcibly taking valued commodities from others. Arnold Buss, speaking from the framework of aggression theory, has labeled this reason for using coercive power as "instrumental aggression."[5] I believe it is fair to say that the goal of gaining by force what would not be freely given has been the reasoning behind most institutional forms of violence. The possibilities of gaining new territory for their own citizens, opening by force new areas of trade, gaining safe ports for their militaries and trading ships, or seizing another nations' metals and precious resources are historically nations' most frequent justifications for war.

Here, then, we have some of the many reasons for the continued development of weapons. For some, the pressure to improve weapons originates from fear, for others from greed, and for still others from the need to promote self-esteem. Thus, the problem of disarmament is plagued by the fact that as one reason for arms building is disputed, many more take its place—all asserting the need for military superiority.

Colonialism and Military Power

Few would argue that the balance of power is altered by the control of superior weapons. In *Pursuit of Power*, William H. McNeil[6] describes military men as macroparasites, who by specializing in violence are able to secure a living without themselves producing the food and other commodities they consume. Thus, a well-equipped and organized armed force making contact with a society not equally well-organized for war acts in much the same way as the germs of a disease-experienced society do. The weaker community, in such an encounter, may suffer heavy loss of life in combat. More often it suffers its principal losses from exposure to economic invasions that are made possible by the military superiority of the stronger people.

Each new improvement in weapon systems that allows one nation temporary power over other nations causes far-reaching changes in the social and political structure of a society. This is because the control of military power underlies the governance of most societies throughout history. We can see this ability to control when we examine the colonial expansion of European powers into India, Africa, and the Pacific in the late 18th, 19th, and 20th centuries.

During this time, even small detachments of troops, equipped with up-to-date European weapons of small arms, ship's canon, and artillery could defeat Africans, Asians, and natives of the Pacific islands with ease. In India, British traders and military faced poorly equipped and frequently competing local rulers whose armies were no match for Western technology and discipline. For the most part, explorations into Africa and the Pacific encountered native populations equipped with primitive weapons of rocks, spears, and bows.

Captain Cook's description[7] of an armed encounter with South Sea islanders typifies the European use of superior weapons to control natives. After seizing as hostages the chief of the island of Raitea and the chief's son, daughter, and daughter-in-law, Captain Cook to his surprise finds that the natives in turn have taken two of his principle officers as prisoners.

I was told that a party of natives had seized Captain Clerke and Mr. Gore. I instantly ordered the people to arm and in less than five minutes a strong party was sent to rescue our gentlemen. [Only] two or three muskets were fired to stop the canoe. To that firing Messrs. Clerke and Gore owed their safety, for at that very instant a party of natives, armed with clubs was advancing toward them. On hearing the report of the muskets, they disappeared. (p. 50)

In the same vein, John Ledyard, a marine on Captain Cook's last voyage, narrates in his journal[9] how guns, Western technology, and perhaps bluff allowed the ship's company to survive safely among 10,000 natives on the Island of Togatapu in the Friendly Islands. Ledyard writes:

These exhibitions of club-fighting on the part of the natives were considered by us of dubious light . . . for we were not certain they were solely intended to entertain us and we always had the guard under arm. The spectators on these occasions amounted to above 10,000 people. However we never let them know . . . that we were jealous of their number or boldness or skill, though we certainly were. Our only defense was our imagined greatness and to accomplish this imagined superiority we resolved to play off some of our fireworks brought off from England. The natives expected musical entertainment . . . and anticipated the satisfaction of finding us inferior to them, but in this they were totally mistaken. For when the first sky-racket [sic] ascended, full one half of several thousand Indians ran off and appeared no more that evening. The Indian chiefs who sat next to Cook and his officers would instantly have worshipped Cook as a being of much superior order and intreated him not to hurt them. (pp. 31–38)

In addition to superior weapons that allowed Europeans to exercise control, the developing technology of the 19th century allowed steamships and railroads to supplement animal pack trains and sailing vessels. Thus, as the logistic problems of supplying armies were solved, obstacles of geography and distance became increasingly trivial. European armies and navies therefore acquired the capacity to bring their resources to bear at will, even in remote and previously impenetrable places.

Colonial Imperatives

The colonial state of mind is based on the core assumption that the colonizer's culture, beliefs, values, and laws are superior to those of the colonized. A corollary of these beliefs is that the colonized are inferior in all ways to those who occupy their land.

Albert Memmi[10] described this state of mind as resulting from simple self-interest by the colonist and the need to justify exploitation of the colonized. He put the matter this way:

> For it was not just a case of intellectualizing, but the choice of an entire way of life. The colonizer, perhaps a warm friend and affectionate father in his native country, will surely be transformed into a conservative, reactionary, or even a colonial fascist. He cannot help but approve discrimination, he will be delighted at police tortures, and will become convinced of the necessity of massacres. Everything will lead him to these new beliefs: his new interests, his professional relations, his family ties, and bonds of friendship formed in the colony. The colonial situation manufactures colonists, just as it manufactures the colonized. (p. 56)

I do not agree with Memmi that simple self-interests, new friends, and professional relations are sufficient reasons to account for the colonist's newly found beliefs. Rather, I believe that traditional colonial attitudes are formed as a result of the drastic imbalance in power produced by military control. That is, the psychological transformation in attitudes that occur in colonists follow a predictable pattern that was described in earlier chapters.

Stages in Colonial Rule

The history of 19th- and 20th-century colonialism suggests three distinct stages of colonial control. In turn, each stage defines the particular attitudes of the colonizers toward those they control.

Stage 1: Conquest

In the first stage, military force is used to overcome native populations. During this time, the occupying country consolidates its military conquest and seeks to impose its own laws, life-styles, and habits. Although colonists control superior weapons, open force and flare-ups of resistance are frequent occurrences. Thus, the balance of power does not completely favor the colonist. As a result, social relations are characterized by a mixture of contempt and distrust of the native populations, combined with a relatively accurate assessment of their ability to resist. This stage describes, for example, American settlers' attitudes toward American Indians in the mid-19th century, English attitudes toward Africans in the early 20th century, and probably Russian attitudes toward citizens of Afghanistan during present times.

Stage 2: Pacification

I will use the concept of identification with the aggressor to describe the changes that occur during this period. In psychology, this concept is used to account for changes in individuals (a) who have been subjected to coercive influence for long periods of time and (b) who finally realize that they lack the ability to resist this influence. For the battered child, the person held hostage, or the timid souls threatened by bullies, it is simply a matter of survival to progress from the belief "If I do whatever they want, I will not be hurt," to "I deserve what is happening because I am less worthy," to "I will help them in all ways because their enemies are my enemies."

I believe that this process of identification with the aggressor can be applied to describe the experience of people who are militarily occupied for long periods by technologically superior countries. In this stage of colonialism, the occupied country has been pacified and foreign rule established. Under these circumstances, the indigenous populations frequently discard their own culture and accept the habits, values, and laws of their conquerors. That is, they give up their own criteria for defining personal worth, identity, normative values, and collective culture. Instead, the values and culture of the colonial settlers and administrators are substituted. Military power, while always present, moves to the background, as successive generations of the native populations voluntarily do what the colonists want them to do. Control moves from threats and intimidation to simple suggestions and requests. When such changes occur, the discrepancy between the native population's own culture and the colonist's culture fosters the view that the former is less worthy than the colonizer's. The attitudes of colonizers during this period are perhaps best described as demanding and condescending, but certainly not distrusting of the loyalty and devotion of the native population.

It is in this stage that one finds complete expression of the metamorphic effects of power. The occupying administrators and settlers are respected and deferred to by the now servile natives. On the island of Tahiti, for example, this process of dominance and submission occurred within 50 years of the Tahitians' first contact with Europeans.[10] During that time, the combined weight of Western technology, values, and military control overwhelmed the 1000-year-old Tahitian culture. This occurred for several reasons. First, the Tahitians were unable to resist Western muskets and cannons. In addition, there were three European commodities that the Tahitians wanted: one was iron, the second was muskets, and the third was rum. The introduction of each commodity further disrupted Tahitian culture, and at the same time increased the Tahitians' dependency on Europeans. The result was that in a short time the Tahitians accepted the culture and habits of Western society and voluntarily complied first with British administrators and missionaries and then with their French counterparts.

Tahiti was first visited by the English explorer Wallis in 1767, then by the French and Spanish over the next several years, and several times in the 1770s by Captain Cook, who established the island's longitude using newly invented ways of recording time accurately. The Tahitians had a primitive society with a well-established religion and family life, an economy based on the cultivation of crops and fishing, and a well-established hierarchy of power vested in several chiefs, who carried out occasional warfare with well-defined rules concerning killing. Within 30 years of its discovery, the Tahitian population had been reduced—through European diseases such as measles, smallpox, diphtheria, tuberculosis, and venereal disease—from well over 200,000 to perhaps 20,000. During the next 25 years, while the missionaries were active, the numbers continued to fall until there were no more than 6,000 to 7,000 Tahitians left. During this time, the internal structure of the society was weakened by the introduction of Western arms. That is, small bands of armed deserters from whaling ships introduced European forms of warfare that eventually destroyed the traditional authority of the Tahitian chiefs. Tahitian culture was further dismembered by English and then French administrators, who introduced the idea of money in exchange for work, laws that defined what was and was not criminal behavior, prisons, and alcohol. English missionaries intent on introducing Christianity continued the destruction of Tahitian culture by requiring "suitable" dress for the women, prayer, church attendance, prohibitions against dancing, and the abolishment of the Tahitians' own religion as barbaric. Feeble attempts at resistance, including armed revolt, were cut down by Western weapons, which, early on, demonstrated superiority over stones and crude spears. Faced with this onslaught of superior technology, the Tahitian society collapsed completely within 40 to 50 years. Today, it has been replaced by a native society that has long forgotten its own culture, and whose day-to-day life is controlled by foreign rule.[10]

Stage 3: Withdrawal

In the last stage of colonialism, the occupying country's ability to maintain military superiority is weakened for various reasons. These include rebellions and guerilla warfare, and/or internal political or financial turmoil within the occupying country. But for whatever the reason, the occupying country finds itself forced to withdraw its military force.

Since World War II we have seen this process of disengagement by Western military forces from Africa, India, and parts of Asia. As the colonists' power base erodes, social relations also change between the remaining colonizers and the native population. That is, colonists are forced to redefine their relations with the natives, who now have equal or greater power than the colonists.

For example, changes in evaluations of the Indians by the British were reported after World War II as Britain found it could not fend off Indians' demands that they withdraw their troops. Thus, as one response to growing demands for independence, Britain established an Indian equivalent of its officer training school, Sandhurst. For the first time it was possible for British officers to meet young middle class Indians on equal terms. "I found these young men fascinating," recalled Reginald Savory, one of its first instructors, "[To my surprise] they were very outspoken, highly intelligent, and highly critical. They told us 'You don't know India. All you know are your servants and your sepoys'" (p 202).[2]

A similar upgrading of Indian abilities was reported by Harley James, an English administrator responsible for overseeing a sensible transition from English to Indian rule: "The astonishing transformation was after Independence when it was discovered that the young Indian was prepared and able to do all the things that one expected of a young Britisher. (p. 211)."[2]

The Metamorphic Model of Power

In the remainder of this chapter I will examine through the words of colonists and soldiers and explorers their social relations with native populations. As in the previous chapters, the metamorphic model of power will serve to organize the discussion. In particular I will examine how the nearly absolute control of military superiority sets the stage for establishing the settlers' evaluations of the colonized, the kinds of social relations that are formed, the settlers' evaluations of themselves, and changes in settlers' moral standards.

The accounts of this chapter are taken mainly from the published diaries and interviews with British colonists in India, Africa, and the South Seas during the late 18th century through the 20th century. I have chosen this period simply because of the availability of diaries, interviews, and public documents that describe the experience of colonizers.

One may ask why I do not use accounts of soldiers and administrators currently involved in military violence against unarmed or poorly armed people. Certainly accounts of the use of superior technology to overwhelm native cultures are found with numbing frequency in today's daily newspapers. As I write, the Ethiopians employ violence against helpless hordes of Eritreans, Soviet armies only recently ceased employing violence against the Afghans, and the People's Republic of China employs violence against the people of Tibet. In this last instance, a deliberate massive resettlement program has attracted millions of Chinese settlers into Tibet, where they enjoy unusually high wages and good living conditions. In contrast, the native Tibetan has far poorer health, a lower literacy rate, more primitive housing, and a life expectancy 20 years below the Chinese average.[11] Thus, colonialism prospers today as it has throughout history. What is not available, however, are the occupying forces' descriptions of their day-to-day lives and social relations with those they now control. This is not surprising. As I mentioned earlier, people in positions of power resist being studied.

The same problem exists in earlier accounts of colonization by historians during the Greek and Roman eras such as Plutarch and Josephus. These historians provide almost endless descriptions of armies marching, dynastic wars, subject peoples in revolt, and the success or failure of these revolts. We are given, however, few accounts of the relations between rulers and those they ruled, nor are we told if the ruler's views of themselves were transformed as a result of their ability to dominate. Yet these accounts of day-to-day lives define colonial attitudes and behaviors.

Attributions of Control

In previous chapters, I described how the control of people through technology resulted in derogation of the less powerful. Since military technology allows total control, we can expect the resulting transformations of attitudes and values by colonists to be far more extreme than those of managers and physicians. The general rule is that total control produces total contempt.

I will begin by examining various causes, both practical and psychological, that encourage the colonialist to derogate native populations. The first cause, already alluded to by the psychiatrist Albert Memmi (1965) in his book *The Colonizer and the Colonized*, is that it is easier to influence others if psychological distance is maintained and emotional involvement kept to a minimum.[9] This may be especially true if control is maintained by open or veiled coercion. Thus, to the extent that colonists feel sympathy for the position of the natives they may not want to use coercion. It is psychologically more comfortable to assume that the object of our influence is not as worthy as ourselves. The colonist can then with good conscience carry out acts that he or she would not willingly do with persons of equal power. For instance, Memmi records the following accusations directed by the French at the colonized Algerians, all calculated to minimize the Algerians' worth:

An old French physician told me in confidence with a mixture of surliness and solemnity that the colonized do not know how to breathe; a professor explained to me that "the people here don't know how to walk; they make tiny little steps which don't get them ahead." (p. 67)

Devaluation, however, is not simply the result of expediency, although it plays a part. There is a second, and more basic reason for derogation to occur. That is, that the very act of influencing (using forceful means) causes devaluation of those being influenced.

Let us examine how this process occurs among the colonists. We begin with the colonists' beliefs about control of natives. Not surprisingly, the use of military power increases this sense of control. There is little doubt as to who is in charge when threats are used or underlie even simple requests. In India during the 19th and early 20th centuries, for example, a fair majority of British settlers were in positions where they controlled Indians, although for the most part the British discounted the military basis for their power. Instead, they attributed the reason for their control to what they vaguely described as their "authority." Thus David Dymington, recalling his service in the Indian Civil Service, observed: "We felt ourselves to be a rather superior people. Our superiority was stressed by our authority, whether in office, the tea garden or the court of law, and it was accepted by Indians without question" (p. 184).[4] As Assistant Superintendent of Police, at 19 years old, George Carroll had authority over hundreds of policemen.

In his own mind the question of exercising power never arose because "it seemed so natural that, as an Englishman, I should have power over all my Indian subordinates" (p. 187).[2]

Even the children of the British in India grew up believing that they controlled. "You were always called little master, and you just took it all for granted" (p. 12).[2] George Carrol remembers as a boy kicking his servant, a former Indian soldier, just to show he was master.[2]

As adults, expressions of dominance and submission were day-to-day events. All believed in white superiority. Thus, if an Indian came along while a white person was riding along a mountain track, the Indian dismounted from his mule or pony and got out of the way. Similarly, an Indian carrying an open umbrella was supposed to shut it when a white person passed. Incomprehensible customs? Perhaps not, since in all ways these signs of deference acknowledge the absolute control of the colonist and the acceptance of this control by the colonized.

The Psychological Consequences of Control

The psychological consequences of perceiving oneself in control of another person's behavior are the same whether such control involves a dominant husband and a submissive wife, or a colonist and a servile native. In all instances, the perception that the person being influenced is not in charge of his or her own behavior causes other people's evaluations of that person to change for the worse. The reason for these unfavorable evaluations is that controlled individuals are seen as not responsible for their own behavior. In the instance of the colonist, the native's behavior, no matter how excellent, is seen as guided by the colonists' orders rather than by the abilities and motivations of the native. Hence, the native is not given full credit for anything he or she does.

Applying these ideas to the use of military power suggests that the more absolute the potential for such control, the greater the derogation of the native. Contrast, for example, John Ledyard's descriptions of the defenseless aborigines of Tasmania with the fierce Maori of New Zealand.[8]

The new Hollander (Tasmanian aboriginal) is a mere savage, nay more, he possesses the lowest rank even in this class of beings. They are the only people who are known to go with their persons entirely nakid . . . They have neither weapons of defense or any other species of instruments applicable to any other of the various purposes of life . . . They have no canoes and they have no houses to retire to . . . They appear also to be inactive, indolent, and unaffected with the last appearance of curiosity . . . Their features discordant and without any kind of ornament or dress. (p. 12)

Being perceived as less than human, it is not surprising that the Tasmanian aborigine was for all practical purposes eliminated from Australian culture within 100 years of the discovery of this land by Cook.

The fighting ability of the New Zealand native produced a different appraisal by Ledyard.

The New Zealanders are generally well made, strong and robust, particularly their chiefs, who among all the savage sons of war I ever saw are the most formidable. It is their native courage, their great personal prowess, their ineversible intrepidity, and determined fixed perseverance that is productive of those obstinate attacks we have found among them when we have appealed to the decisions of war. Among all the savage sons of war I ever saw, they are the most formidable. When a New Zealander stands forth and brandishes his spear, the subsequent idea is: There stands a man. (p. 15)

Devaluation of natives, then, grows as the potential for control exists. Concerned with the growing riots in India and in Palestine in 1921, and these countries' demands for freedom from British rule, Winston Churchill told Parliament:

In the African colonies you have a docile tractable population who only require to be wisely treated to develop great economic capacity and utility; whereas the regions of the Middle east and India are unduly stocked with peppery, pugnacious, proud politicians, who happen at the same time to be extremely well armed and extremely hard up. (p. 221)[12]

Thus, where control is seemingly well-established, the adjectives applied to describe the native population differ from those applied to native populations in revolt.

As control is consolidated, we find that devaluation of natives occurs along many dimensions. Thus, in East and Central Africa there were expressions of disgust and repulsion at native eating habits, food, dress, intelligence, and even natives' ability to express normal Western sentiments of love and family life. In India, Terrence Milligan[2] could say in recalling his childhood days in military cantonments: "I never thought anything else except that the Indians were inferior to us. When I first went to Calcutta, you could walk down Chowringee [Street] and the Indians walking in the opposite direction would just get out of your way." (p. 215)

In India, Moslem and Hindu cultures were viewed by many English colonists as barbaric and crude. In the late 19th century there was little study of Hindu art. Rosalie Roberts,[2] the wife of a British administrator, was interviewed about her impression of India. She recalls how she once visited a local temple, very old, with hundreds of carvings, and on every ledge a tiny lamp. She found it particularly repellent: "All the little niches had idols, where there had been sacrifices and blood spilt" (p. 211).

In 1845, the directors of the Missionary Society, boasting of their success in establishing Christianity as the religion of Tahiti, wrote to the Missionary Board in England that the Tahitians, before the missionaries came, had been "mean, cruel, malignant, dishonest, untruthful, depraved, ferocious, quarrelsome and warlike."[10] Furthermore, the Tahitians were "idolaters sunk to the lowest possible depth of moral degradation, whose system of superstition was one of the most absurd and sanguinary that ever prevailed among mankind" (p. 172).

Note that 45 years earlier, when the Tahitians' tribal chiefs still ruled and Western presence was found only in occasional visits by American whalers and British Navy and trading ships, the same missionaries had a more favorable, if still condescending opinion of the Tahitians. In 1797, they wrote to the Missionary Board in England that the Tahitians were "good people, generous, kind and happy." (p. 175)

In short, for the colonists, as their control increases and the natives become docile, predictable changes occur in their appraisals of the natives. In all instances, as power increases so, too, does denigration.

Moving Away from the Powerless

There is nothing surprising in the statement that we avoid the company of those we do not like. Perhaps it is somewhat more surprising that we also distance ourselves from those we control. Instead of disdain, it is equally logical to assume that we should be grateful for the many services that they provide. However, this does not happen very often. Indeed, diaries and public records suggest that colonists expend considerable energy moving away physically, socially, and psychologically from those they control.

The metamorphic model provides several reasons for this social distancing, in addition to the simple reason that colonists dislike the colonized. First, several social scientists have observed that people in high power can afford to be indifferent to those of lesser power, since the less powerful are neither a threat to their security nor in a position to offer any insights of interest.[13] If one assumes there is little to be learned from the native culture, then why bother spending time with natives?

Another reason for movement away stems from distrust of the motives of the less powerful. Social philosopher Ronald Sampson,[14] has observed that persons in position of power are often repelled by the obsequiousness of the less powerful, their lack of candor, and their penchant for flattery. Thus, the colonialist can never be sure that there are not ulterior motives in any positive statements expressed by the native. These suspicions also contribute to a preference for social distance.

A final explanation for the preference of the colonist for social distance comes from the suggestions of B.F. Skinner in his book, *Beyond Freedom and Dignity*, that Western society values free choice. By extension, we may also desire to avoid persons who appear not to be in control of their own behavior, but in fact are controlled by us. Fundamentally, I would argue that it is natives' lack of freedom of choice that provides the motive to move away from them. To be controlled by another person robs one of those very qualities of dignity and self-worth that attract one person to another.

The accounts of colonists are filled with their many attempts to avoid contact with the colonized. At a social level, there were explicit norms established in India. "We didn't mix with the Indians at all," remembers Mrs. Norie, who returned to India in 1893. "Army officers did not associate with Indians of any class other than the servant class, to whom they just gave orders . . . They were regarded as a subject race." (p. 207).[2]

As servants, Indians were almost invisible, even when in the most personal of confrontations, as Radclyffe Sidebottom tells it:

My wife would have the bath first and the ayah would dress her. I would go in and have my bath and my personal servant would bring in a drink and my wife and I would carry on a conversation as if the two servants in the room weren't there. (p. 77)[2]

These attitudes for the most part were held by men and women, who at home in Britain shared in the universal beliefs of the benefits of democracy and equality. Yet as administrators and settlers, the ability to control almost automatically transformed the value of democracy. John Morris recalled in an interview that in the 1920s he was on a train in Bombay and discovered that the other berth in his room was occupied by an Indian. "I am ashamed and sorry to say that by that time I had become affected by the mentality of the ruling class in India. I said to the stationmaster: I want this gentleman ejected " (p. 199).[2]

And then there is the matter of social contacts. People who study social power consistently find that the more powerful prefer to mix with those of equal or greater power than themselves. The whole basis for social chitchat is distorted when talking to the less powerful. As one English colonialist observed while talking about the English-only rule in social clubs: "We were always watching our step and watching what we said and there's a certain relief to go amongst people of our own race and let our hair hang down" (p. 78).[2] Thus, it is not surprising that by custom and by local law, social exchanges between the Indians and the British were frowned on. If a soldier was seen as joking or talking to an Indian, especially the same Indian two or three times, Charles Adams reported, "he was jeered at and called a 'white nigger' " (p. 79).[2]

Among the middle and upper classes of settlers of India, the club was the place for getting together, dances, dinners, sports, games, and exercise. Every station had its club. The bigger ones had tennis courts, squash courts, and even at times a golf course. Reginald Savory, an older settler of India regretted that clubs would not allow Indians to be members.[2] This was frequently written down in their constitutions. H.T. Wickham described the controversy that arose in his club when the issue arose in 1921, a time when Indians were strongly pressing for equality and independence from Britain. "There was a question of allowing Indians to join the club. A large number of the members didn't like it. Their chief objection was the fact that the Indians, if they joined the club, would consort with the female members" (p. 99).[2]

And so in the home, on the street, at work, and in the clubs, the colonizers recreate their own world and shut out the colonized. The very architecture of the colonizers was designed to express their separation from the colonized. In Africa and India, the homes and clubs that were constructed emphasized the great distance between the cultures. While the goal for the settler was to make the country as homelike as possible, it also served to emphasize the great gaps between the English and the natives. Imagine if the Kikuyu tribe of Kenya occupied New York City, destroyed its buildings, and constructed bush houses throughout. The very strangeness would keep New Yorkers away.

Power Raises Self-Esteem

Not only does control increase the colonists' contempt for those they control, but in many instances control raises the colonists' self-esteem. These positive changes are most likely to occur during the second stage of colonization, for it is then that the native population identifies with the colonists and military control moves to the background.

There are several ways in which the colonist's self-regard is raised as a result of being in charge. Most obviously, power allows individuals to lead a comfortable, if not luxurious, life. *The Settlers Guide: New Homes Under The Old Flag*, published in England in 1902, informed potential colonists of the many rich lands awaiting them under the British flag. From Fiji to Borneo to Africa there were lands to be worked at pennies per acre and a ready supply of natives or Indian coolies for labor at wages of 1 shilling per day without rations. If properly controlling, one could live like a king, surrounded by hardworking and docile natives. Surely living in this kind of luxury must elevate for the better colonists' views of themselves.[12]

Beyond the material comforts that accrue to colonists, there were a wide range of psychological comforts that came from having power. Thus, colonists during the second stage of colonialism frequently were the recipients of flattery and well-wishing from the natives anxious to receive favors. To the extent that this flattery was believed, it was difficult to not think of oneself as something special. As George Henry Mead[15] pointed out many decades ago, the definition of self is strongly defined by the attitudes of others. The more favorably others act toward you, the more favorable your opinions about who you are.

Rosamund Lawrence described the pomp of life in an Indian station:

When we got there the band was playing and banners saying: Welcome .... They brought garlands of roses and put them around our necks. Then we got into a four wheeled bullock cart painted white with tigers and leopards painted inside and drawn by great big bullocks with garlands of roses round their necks. We just sat there with rows of people lining the road saying 'Salaam, Salaam.' I felt like Queen Alexandria driving through Hyde Park (p. 65).[2]

"Whenever my father came in, recounted Iris Portal, her father's bearer Gokhal always bent down and dusted his shoes with a duster on the steps" (p. 83).[2] Such feudal relations seemed strange to colonizers when they first arrived. Norah Bowder found to her horror that her husband was being dressed and undressed by his bearer, even to the point of having his bath water poured over him. "I found this extraordinary because when I first met him in England he was doing everything quite competently" (p. 68).[2]

Not only were the colonists treated with exaggerated deference, but their children also were. In India, even the smallest of children lived in a land where their needs were cared for by servants. "I grew up in bright sunshine, I grew up with animals, I grew up believing that white people were superior," recalled the son of a British army corporal. (p. 1).[2] Another former Indian settler, John Rivett-

Carnac, fondly recalled his childhood: "There were many servants who treated me as an adult, with the same respect as they treated my father. The result was that I got a very great idea of my own importance" (p. 12).[2]

Deference, flattery, and servility cannot help but raise the colonist's self-esteem, if only in comparison to the natives with whom he or she lives. However, there were more than endless servants, servility, and flattery to raise the colonists' self-esteem. Because of the power they controlled, the colonists found that their ideas and opinions and suggestions were readily agreed with. Common sense told the native that it would be costly to disagree. Even the most foolish of suggestions may be carried out by natives who want to keep in the good graces of the colonists. This public compliance frequently led colonists to believe that their ideas and views were superior to those held by others, when in fact compliance was not based on the superiority of their ideas, but on the superiority of their power.

Leonard Woolf, himself a writer and the husband of novelist Virginia Woolf, went as a young man from Cambridge to Ceylon in 1905 to serve as an administrator of British rule.[12] He was sent to a vast Pearl Fishery camp at Marichchukaddi where he found it quite natural "to keep order over 40,000 Arabs and Tamil divers with a loud voice and a walking stick." Later he was sent to an arid sandy peninsula called Mannar, which had jungle villages, and where he was the only white man. He wrote: "Out in the jungle it is different. You are still absolute; the villagers still come to me to settle disputes and my word is law." Woolf happily testified that his colonial experiences "made a man of him" (p. 54).

Thus, power and unbounded control of others can change the individual colonist's opinions about himself or herself far more rapidly and painlessly than even the best forms of psychotherapy, and these beneficial changes can occur in a far shorter period of time than can be accomplished through therapy. Flattery, deference, compliance, and the unchallenged use of authority all serve to enhance the colonizer's self-image.

Doing Dirty Work: Changes in Moral Values

It has been continually observed in the social science literature that powerful persons evolve new codes of ethics that serve to justify their use of power. History and literature contain endless accounts of acts by the powerful that violate common standards of decency. Thus, for example in Sophocles' drama, *Antigone*, when this noble heroine invoked the universal laws of the gods as justification for the need to give her brother a proper burial, King Creon countered with a new set of laws, the chief of which was reverence for his laws and commands.

The diaries of colonists, as well as even a casual reading of the daily newspapers, report a never-ending stream of accounts of colonists' dual standards of morality when dealing with native populations. I am not necessarily talking about colonists conducting gross acts of torture and murder, although these occur often enough, but rather the general belief that it is acceptable to behave in unacceptable ways when having commerce with natives.

Imagine, for example, you and your family being invited to a dinner party as honored guests, being treated royally, and then when the dinner was over, discovering that your family cannot leave. Rather, they are held by your host as prisoners. If this happened in any known part of the world, the behavior would be described by all as legally and morally wrong. Not so, however, if done by Captain Cook[7] aboard the British sailing ship HMS Resolution in the year 1777, and the guests were a family of natives. Then the imprisonment would be described as a clever strategic move to control the natives. As Cook writes:

On the 24th I was informed that a midshipman and a seaman belonging to the [sister ship] Discovery were missing. As these two were not the only persons on the ships who wished to end their days at these favorite islands [Raiatea and Bora Bora], it was necessary to get them back at all events in order to put a stop to further desertion. I determined to have recourse to a measure which would oblige the natives to bring them back.

Soon after I invited the Chief Oreo, his son, daughter and son-in-law on board the Resolution. After food I resolved to detain the three last until the two deserters should be brought back. With this view in mind they were confined. The Chief began to have apprehensions . . . . I soon made him easy by telling him that he was at liberty to leave the ship and to bring our men back. If he succeeded his friends would be delivered up; if not, I was determined to carry them away with me. The Chief's daughter's [cries] resounded from every quarter, and the women [in surrounding canoes] seemed to vie with each other mourning her fate. (p. 167)

The strategy worked and the deserters were returned. Cook used this same strategy – intimidation and violation of trust as I think of it – on at least three different occasions. In all instances, native chiefs were invited to meals and then, to their surprise, held hostage. Indeed, Cook was killed the next year by native Hawaiians while he was attempting to lure their king aboard his ship to be held as hostage. In this instance, however, so much anger had built up between the Hawaiians and the British that when the Hawaiians saw Cook leading their king toward the shore to board ship, the Hawaiians detained the king by force. The anger between the British and the Hawaiians was caused, on the one hand, by the arrogance, frequent brutality, and indifference to native custom of the British, and on the other hand, by the Hawaiians' frequent attempts to steal metals and other products of Western technology. What I find curiously fitting in the manner of Cook's death is that he was murdered by a native who used an iron dagger traded to him by Cook in exchange for food and supplies for his ships. In this instance, at least, the traditional colonial exchange of native foods and unprocessed raw materials for the colonists' manufactured goods backfired badly.

Thus, threats to power cause changes in normative values. Not surprisingly, when military power is used, changes in values are such as to justify killing natives. Such changes occur most often during the first and third stages of colonial occupation when soldiers and settlers are either attempting to seize territory from its occupants or using force to overcome native uprising.

The diary of the British soldier Richard Meinertzhagen,[12] written during the first stage of Britain's colonization of wide areas of Africa, illustrates these new sets of values. His diary records his efforts to pacify native populations, mainly, it would seem, by killing them. He came to East Africa in 1902 to join the King's African Rifles. According to his biographer, Valerie Pakenham, his code was a mixture of Social Darwinism (the nation's right to impose itself on lesser breeds) and a belief in the moral benefits of British rule. He wrote on his first posting away from Nairobi, in Kikuyu country:

Here we are, three white men in the heart of Africa with twenty nigger soldiers, and fifty nigger police administering a district inhabited by half a million well armed savages who have only recently come in touch with the white man. The position is most humorous.

His first action was against a Kikuyu village.

I strongly advised immediate and drastic action in the shape of surprising the village that very night.

After a dawn raid they found the inhabitants dancing in full war paint.

There was a rush . . . into the surrounding bush and we killed about 17 niggers. Two police and one of my men were killed. I narrowly escaped a spear which whizzed past my ear. Then the fun began. We at once burned the village and captured the sheep and goats. After that we cleared the valley in which the village was situated and killed a few more niggers, who finally cleared off. (p. 198)[12]

Not much of your British fair play, laws, and Christian theology here. Such moral standards wait for pacification.

It is only in the second stage of colonialism that an approximation of the colonists' own social and moral values can be safely substituted as guides for commerce with natives. Then, since military force is not needed to control, colonists feel comfortable in substituting their own laws and religious systems to judge the native. During this stage, colonists frequently mention their felt obligations to bring civilization to the natives. Usually this meant better sanitation, Western schooling, religion, and the substitution of Western technology for the natives' less developed ways of coping with uncertainty. Once the natives have been pacified, settlers feel obliged to be guided by shared normative values. Thus, the natives can obtain justice by appealing to the same value system as the settler. "Be very fair to your servants," Olivia Hamilton was told when arriving in India.[2] The same attitudes allowed the despised Indians to make claims in courts of law and in day-to-day transactions with the British. No longer could the settler take by force of arms whatever was wanted from the natives.

Dual standards again emerge as the colonists must resort to force to defend what has been seized. Thus, the British General Dyer in 1919 in defending British control of India found moral justification for using superior military force.[12] He brought his troops to face a large, but unarmed crowd of Indians at Amritsar, who were demonstrating for Indian independence from Britain. He ordered his troops to open fire, killing 375 people and wounding 1,500. The general later testified that he brought his troops to the area for the deliberate intention of shooting. He wanted to "create a sufficient moral effect from a military point of view." Thus, from the settlers' point of view the general's actions were not from simple expediency of preserving power, but were in fact grounded in the moral need to teach the natives for their own good.

A Final Account

I have personalized the account of technology in this chapter by using the words and recollections of colonists. For the most part these recollections make no mention of weapons or coercion except in the beginning and final stages of colonization. But the reader should be aware that these words and recollections could only have been made because of the settlers' control of power based on a technology of weapons. In this instance, Western gains in military technology allowed the French, German, and British armies in the 19th and 20th centuries to overwhelm the primitive defenses of peoples in Africa, Asia, and the Pacific islands.

The colonist's state of mind, then, draws its inspiration from the same source of technology as does the physician's relations with patients and the industrialist's relations with workers. In all instances the added edge provided by technology fosters the perception that people—patients, workers, and the colonized—are not in control of their own behavior, but are controlled by outside forces. In turn, such perceptions almost automatically encourage the development of derogation of the less powerful, social distancing, and the evolution of separate moral codes to justify relations with the less powerful.

There are, of course, differences between technologies in terms of their effects upon powerholders. This chapter has moved the examination of technology from indirect to direct control of people using force and intimidation. Unlike the physician or the manager, the user of military power cannot ignore the fact that he or she controls other people's behavior and threatens harm. Thus, the metamorphic effects of power are intensified in terms of derogation and social distancing when coercion is involved.

References

1. Osgood, R. (1976). *Force, order, and justice*. Baltimore: Johns Hopkins Press.
2. Allen, C. (1985). *Plain tales from the raj*. New York: Holt Rinehart and Winston.
3. Freud, S. (1957). *Civilization and its discontents*. London: Hogarth Press.
4. Fanon, F. (1963). *The wretched of the earth*. New York: Grove Press.
5. Buss, A.H. (1971). Aggression pays. In J.L. Singer, (Ed.), *The control of aggression and violence*. New York: Academic Press.
6. McNeil, W.H. (1982). *The pursuit of power*. Chicago: University of Chicago Press.
7. Dale, P.W. (1969). *Seventy north to fifty south: Captain Cook's last voyage. A modern annotation of Cook's last journal*. Englewood Cliffs, NJ: Prentice-Hall.
8. Hitchings, S.H. (1963). *John Ledyard's journal of Captain Cook's last voyage*. Oregon: Oregon State University Press.
9. Memmi, A. (1965). *The colonizer and the colonized*. Boston: Beacon Press.
10. Howarth, D. (1983). *Tahiti: a paradise lost*. New York: Viking Press.

11. Kirkpatrick. J. Communist Aggression. *Philadelphia Inquirer*, December 14, 1987.
12. Pakenham, V. (1985). *Out in the noonday sun: Edwardians in the tropics*. New York: Random House.
13. Zander, A., Cohen, A.R., & Stotland, E. (1959). Power and the relations among the professions. In D. Cartwright (Ed.), *Studies in social power*. Ann Arbor, MI: Institute For Social Research.
14. Sampson, R.V. (1965). *Equality and power*. London: Heineman.
15. Mead, G.H. (1934). *Mind, self, and society*. Chicago: University of Chicago Press.

CHAPTER 8

Solutions

*All scientific speculations, of whatever kind, as human endeavors, must per-
force be subordinated to the idea of progress, to the true general theory of
human development* (Auguste Comte)

During the Enlightenment, few social philosophers seriously wondered what
society might look like once the use of reason had succeeded in making life better
for people. It was simply assumed that, in such a world, God's promise that man
would have dominion over the birds and the beasts and the seas and the dry lands
would be realized, and that a world in which its citizens consciously choose to
base their lives on reason would develop all that is best in humanity. In their
visions of future worlds, Plato, Thomas More, and Auguste Comte took as their
theme the intellectual growth and responsible participation of man in society.
They envisioned a society that empowered people with the ability to think and
choose for themselves. A society in which "the spirit of the law was steeped in
man's nature like a dye,"[1] and "where each person contributed their skills to society, but not under compulsion."[2]

But what of today's Utopians? The theme has shifted from a society based on
reason to one based on control. Books such as Skinner's *Walden Two*, Orwell's
1984, and Huxley's Brave New World, and films such as Fritz Lang's *Metropolis*
and Woody Allen's *Sleeper*, describe a future in which masters employ technology
to control the ordinary person. Sometimes the control is applied for the person's
own good, as in *Walden Two*. But in most instances the Hobbesian assumption is
made that the ordinary person is better off without choice.

We find, then, that the autonomous and rational human of earlier visions has
been replaced. Today's scenarios of the future see humans controlled by
machines, drugs, and coercion. Gone is the picture of responsible citizens, each
voluntarily contributing to society. These citizens are replaced today by machine
tenders doing mindless work without satisfaction and without thought about their
roles in the state. Thus, the Enlightenment's vision of human beings' potential to
grow intellectually and to exercise choice and reason has been replaced in today's
utopias with the vision of people controlled by technological forces. One could
say that the vision of the future has soured.

The notion, then, of mastery and control appear closely linked in any discussion of modern technology. These modern visions of the future are far from the promises of technology that were described by Florman[3] and Weinberg[4] in the opening pages of Chapter 1. Technology may generate new opportunities, but it also is seen as able to unleash the worst elements in human nature. What remains puzzling in these modern visions is that we are not told why the masters act so cruelly, or why they behave with such indifference to the fate of those they control. The logic is simply that cruel taskmasters gained power.

In this book, I have presented a psychological explanation for this cruel behavior based on the assumption that the exercise of power transforms values and behavior. The major theme is that technology consolidates power and reduces individual autonomy. In turn, as controllers of technology perceive that workers and citizens are not in charge of their own lives, they take on attitudes of derogation and indifference toward those workers and citizens. Chapters 4 through 7 presented evidence in support of this view. People seem unable to use in a restrained way the power provided by technology—and that is a serious problem. In this chapter, I examine the extent to which this psychological problem has been considered in discussions of either the social problems generated by technology or the proposed solutions to these problems.

Problems of Technology

Concerns about the use of technology are not new. David Dickson[5] pointed out that throughout history man has been warned that he was creating forces he would be unable to control, that machines would eventually take over the planet and demand total obedience from humans, and that to place one's faith in science and technology was to make a pact, like Faust, with the devil. Yet when we compare the range of concerns about the use of technology that are expressed today to those expressed, say, 250 years ago, we find that our problems have increased many times. The problems caused by technology today are not only about losing our souls to the devil. Today, there is a strong possibility that we may lose our lives as well.

In more detail, some writers focus on environmental problems such as environmental pollution; the depletion of our natural resources; the problems of nuclear accidents; and the ways in which the toxic residues of industrial processes are spreading over the seas, over the land, and in the air. Others are concerned with problems caused by technology at the workplace, such as routinized jobs, alienation of workers, and the use of bureaucratic and hierarchical controls to maintain satisfactory levels of employee performance. Still others write about the ways in which technology tends to create work that isolates people from each other and from the community, and to encourage passive modes of adaptation, such as watching television and video tapes. The problem here is that isolated and passive individuals tend to be powerless and less able to resist

TABLE 8.1. Problems caused by technology.

A. Environmental
   a. Pollution, depletion of resources, toxic wastes[11]
   b. Dehumanized cities, overcrowding, traffic[6,11]
   c. Creation of environments that are inconsistent with the human needs; the denial of
      women's values and imposition of male-oriented values of predictability, repeatability,
      quantification[12,13]
B. Work
   a. Creation of low-skilled jobs, routinized work[14,15,16,18]
   b. Loss of decision-making power by workers[18]
   c. Worker alienation[16,17]
   d. Bureaucratic and hierarchical control[17,19]
   e. Exploitation of workers[18]
   f. Work has become reactive rather than proactive; for example, passive monitoring[8]
C. Society at large
   a. Technology concentrates power[19]
   b. Erosion of democracy[20]
   c. Reduction of individual power by encouraging isolation and passivity[34]
   d. Social stratification: scientific elites versus the uninformed masses; those capable of tech-
      nological training versus the unemployable[14,21,32]
D. Technology controls its own evolution[9,10]

unwelcome attempts to influence them. Still others define the problems of technology as resulting from centralizing power and control. Writers such as John
McDermott,[6] Barry Jones,[7] and Mike Cooley,[8] to name but a few, point out that
the major problem of technology is not technology per se, but how it is used to
sustain and promote the interests of the dominant social groups of society. David
Dickson puts the matter this way in his book *Alternative Technology*[5]:

Technology is a political instrument and becomes an end in itself. Power will move toward
the controllers of technology and away from a poorly informed and increasingly apathetic
electorate. (p. 10)

Finally, there are the ideas of Jacques Ellul,[9] Friedrich Juenger,[10] and John
McDermott,[6] who see little possibility of creating a humane society from technology. Rather, technology is seen as controlling events and giving direction to society, uninfluenced by rational control.

In Table 8.1, I have summarized the major problems noted in the current literature on this subject. We can see in this table, and in this brief review, that the
metamorphic effect of power—the idea that the added power provided by technology may transform, for the worse, those who control it—is not mentioned as a
significant problem. The closest we come to this awareness is the linking of technology to persons who control political and economic power. But here, the idea
seems to be that controllers are by their nature greedy and exploitive, and
perhaps the issue is how to prevent these people from having control, and/or getting more humane people to replace them.

Solutions to Technological Problems

Given the wide range of problems identified with the use of technology, it is not surprising that an equally wide range of solutions has been offered to cope with these problems. As shown in Table 8.2, these solutions seek to redress the problems of technology through technological means, through citizens and/or panels of experts assessing the costs and benefits of a given technology, and through gaining access to political power. All are possible. But it is also apparent that these solutions do not consider the ways in which power may affect the values and attitudes of controllers of technology.

In the next several sections I shall examine the relevance of several of the more important of these proposed solutions in light of our assumptions about the goals, attitudes, and values of controllers of technology. Plausible solutions, in my opinion, should take into account the following assumptions, derived from the metamorphic model:

1. Controllers of technology are concerned with protecting and maximizing their resources.
2. Controllers' definitions of the problems of technology do not coincide very often with the problems as seen by persons who are targets of technological influence, or by impartial scientific experts. The problem for the controllers is how to protect and/or minimize damage to the technology they control. For the public, the focus is on correcting problems of technology that are causing people distress.

TABLE 8.2. Solutions to problems created by technology.

A. Using technology to solve problems
  a. Technological "fixes"[3,4]
  b. Technology will evolve to the point where control is not needed[5,7,17]
  c. Rescaling technology to meet human needs (alternate or humane technologies, job enrichment)[5,7,22,23,24,25]
B. Using people to solve technology's problems
  a. Office of technological assessment[26]
  b. Citizen decision-making groups[24]
  c. Wise decisions by controllers of technology[11]
  d. Panels of experts[27,28]
  e. Educating the public—eliminating technological illiteracy[11,29]
C. Regulating technology through political means
  a. Power-sharing between management and workers (participation, worker committees, worker–management councils)[8,30]
  b. Community protests and agitation[8]
  c. Pressure placed on legislature[5,6]
  d. Maoist solution—requiring power elite to work in routinized jobs[8]
  e. Political revolution to seize control of technology[8]
D. There are no solutions[9,10]

3. Controllers of technology are not particularly concerned with the welfare of
   persons subject to technological influence.

Using Technology to Correct Technological Problems

The solution that appears most acceptable to most people for coping with
problems created by technology is to use more technology.

At its simplest level, the idea is that new technology can fix the problems
created by the old technology. Thus, as I write, refining industries are combating
the loss of ozone in our air by changing the chemical composition of gasoline, and
oil companies in Alaska are employing advanced technology to clean up the
spilled oil that is clogging Alaskan waters. Since the controllers of technology are
the people who devise the solutions, it seems fairly certain that their recommendations will not include doing away with the technology that caused the problem
in the first place, for this might mean the elimination of the use of gasoline to fuel
cars and perhaps do away with pumping oil from the fields in Alaska.

Sometimes we are led to believe that the social sciences can repair the
problems of technology by reducing human selfishness. Thus, for example, B.F.
Skinner, writing in *Beyond Freedom and Dignity*,[31] argues for a technology of
behavior to repair the damage produced by applied physics and biology:

What we need is a technology of behavior. We could solve our problems quickly enough
if we could adjust the growth of the world's populations as precisely as we adjust the course
of a spaceship, or improve agriculture and industry with some of the confidence with
which we accelerate high-energy particles, or move toward a peaceful world with something like the steady progress with which physics has approached absolute zero. (p. 3)

Skinner, of course, assumes that the technology of behavior will shape up the
behavior of ordinary citizens, and leave untouched the benign and humanistic
motives of those who will control the behavior technology. If I am correct about
the metamorphic effects of power, Skinner's well-meaning hopes are unlikely to
be realized. In fact, if such behavioral technologies were developed, they would
be a part of the problem rather than a solution.

Another technological solution to technological problems is based on the
notion that technologies can be created that eliminate pollution, eliminate worker
alienation, and distribute power rather than concentrating it. Advocates for this
approach use design criteria for the development of technology that seek to max-
imize individual worth and self-reliance, that are based on labor-intensive rather
than capital-intensive technology, that use minimum energy to maintain environmental quality, and that decentralize rather than centralize power. At work, advocates of alternative technologies propose that tasks should be enriched and
enlarged—reengineered to allow employees to work more or less autonomously
or in self-managing teams. At the society and community level, the theme is

a countryside dotted with windmills and solar houses, studded with intensively, but
organically-worked plots of land . . . a lifestyle for men and women which involved hard

physical labor, but not over excessively long hours or in a tediously repetitive way . . . and a political system so decentralized and small that individuals – all individuals – could play more than a formal, once every five year, role.[5] (p. 139)

Perhaps the slogan "small is beautiful" adapted from Schumacher's[23] proposals for an alternative technology best describes these solutions.

A problem with these solutions is that they don't tell us very much about how to implement them, aside from Cooley's suggestion about workers seizing control of the country, or Jone's[7] and Dickson's[5] rather vague proposals for putting into power a government committed to dismantling its own current technological structure. It is apparent to me that the installation of alternate technologies requires the active cooperation of persons who own and control current technologies. Yet few, if any, industrial or political leaders have shown any interest in putting such ideas into practice. Lip service may be given to participatory doctrines of management, but as William Faunce[17] points out, managers are attracted to solutions that protect or increase their favorable positions in the power hierarchy. Alternative technologies directly threaten existing power and for this reason alone are unlikely to be voluntarily put in place in industry or in larger society. Usually, when one reads partisan accounts of the establishment of such technologies, one finds them located in obscure regions of the Third World or in small communes where people support themselves with funds provided by outsiders or by selling organically fed chickens, pamphlets, and guided tours.

Reason and Participation

Still another proposed solution to the problems of technology involve citizen groups and "watchdog" committees, who assess the costs and benefits of technology and then make rational decisions about its use. This solution assumes that technology is a relatively neutral force that provides choices[14] and the wisdom of these choices depends on the technological sophistication of those making the choices.[11] A further assumption is that people who experience the costs and benefits of a given technology will participate in decisions about its use. Political essayist Paul Goodman[24] advocates, for example, the model of New England town meetings, where interested parties could meet in government-supported centers to discuss and decide on proposed technological innovations.

Joseph Coates[26] summarizes these kinds of solutions as including: (a) a political system in which elected officials, legislature, and regulatory agencies exercise control of technology; (b) new and better means of anticipating the consequences of technology, through expert opinions and computer modeling of likely future events; and (c) citizens at large and organized interest groups who have access to reliable information and exercise the political process vigorously to reflect their needs and values.

The above is theory. What seems more likely to happen in practice is that the more influential of these watchdog groups represent controllers of technology, rather than people who may be affected by its use. Emmanuel G. Mesthene,[11] former director of the Harvard Program on Technology and Society, has suggested, for instance, that controllers of technology are in the best position to rectify any problems caused by technology—a solution as helpful as asking a wolf to oversee the raising of chickens. Expert advice is provided to the United States Congress by the Office of Technological Assessment, which recommends policy on airports, water pollution, the Alaskan oil pipeline, and the like. For the most part, this federal agency seeks the opinions of all groups involved in any new technology, but gives, I suspect, more weight to the testimony of scientists, technical persons, and/or corporations who have financial interests in the application of technology. As Dorothy Nelkin[32] writes in describing the difficulties of ordinary citizens to affect decisions about technology: "The complaint is that the technical elite are essentially a 'mandarin class,' using their specialized knowledge to serve established institutions of industry." A similar conclusion can be drawn from the research of Dutch psychologist Mauk Mulder[33] who has shown that the opinions of experts are given far more weight when debating a given issue than the opinions of uniformed but concerned citizens.

This is not to say that citizens groups are completely ineffective in shaping technological decisions. Citizens at large—forming temporary interest groups around specific issues such as pollution of beaches, and formally organized groups, such as the Sierra Club and friends of the Earth—have successfully mobilized public opinion. But their successes have been limited. In many instances, the original policies that gave rise to the protests have been reinstated once the public hue and cry has died down. We may also note that citizens lobbies are usually concerned with conserving the environment—preserving green spaces, protecting endangered species, and safely disposing of toxic wastes. Issues such as the deskilling of work and the concentration of power seem not to be a matter of public debate.

Rather, such debate is between management and labor, and, in the United States at least, the evidence suggests that management has won most of the points. Through a process of discussion and negotiation, labor has tried to convince management to allow labor equal say about the use of new technology in the plant and office. The goal of labor has been to limit the introduction of automation, to participate in decisions about how best to use the new techniques, and to protect job security. At best, labor has won agreements to guarantee job security and the retraining of workers who have been displaced by automation. However, attempts of labor to restrict the installation of automated technology appears to have received little success.

Daniel Cornfield's[30] analysis of the introduction of automation in 14 major American industries describes very well, I believe, the difficulties experienced by labor when negotiating with management about these issues. Cornfield found that the most prevalent pattern in these industries was for management to make unilateral decisions to install automation. Traditional labor-intensive occupations have been eliminated over the last 20 years in such industries and occupations as agriculture, the newspaper industry, longshoremen, the postal service, office workers, school teachers, air traffic controllers, the commercial aircraft industry, telecommunications, and more. In these industries, Cornfield reports that labor was not strongly unionized or the economy was expanding so fast that the displaced workers could find new employment. Thus, resistance to the introduction of new techniques was weak.

There were, however, several major industries (auto, steel, coal) in which labor had sufficient power to negotiate formal agreements about the introduction of automated technology. While labor organizations accepted greater automation of their work as part of the bargain, they won agreements to have an equal say with management about its use. We can note that this is a preferred solution by labor to problems of automation. Needless to say, the solution, preferred by unions, to share power is not regarded with great favor by management.

While perhaps predictable in terms of the dynamics of power, it is of interest that these agreements have not worked very well. One of the reasons, according to Cornfield, was management's resentment at having to share decision-making power with labor. To avoid sharing power, Cornfield reports that management has established coproduction ventures with their foreign competitors. These ventures in joint manufacturing have effectively bypassed American labor's attempts to share decision-making power.

What I believe is missing from the proposed solutions that I have reviewed is an account of the very real divergence in goals, interests, and values between persons who control technology and those who are controlled by it. I argue that the psychology of persons who control must in some way be taken into account before workable solutions can be reached. Unfortunately, I don't know how this rapprochement is to be achieved. One can argue, as do the more militant critics of technology, that the solution lies in replacing persons who control power with new faces from the ranks of labor and citizens groups. This could be done, perhaps, through revolution or by replacing controllers every 3 to 4 years.[8] Besides being a foolish recommendation, considering Chairman Mao's disastrous attempts in China to randomly distribute positions of power, this kind of rotation does not take into account the likely transformations in values and attitudes of these new heads as they gain experience with their newfound power. One can expect them to become as disdainful of the masses as the controllers who preceded them.

Perhaps the evolutionary solution described by William Faunce,[17] in which the role of controller tends to disappear, may be the answer. Based on Daniel Bell's description of the postindustrial society, Faunce speculates that automation contains the seeds for the destruction of industrialism. This will happen because a new breed of professional, technical, and skilled workers will be needed by industry. These employees work best when bureaucratic and hierarchical controls are removed. Such workers are self-motivated and intrinsically interested in doing good work. As the nature of work changes, Faunce argues, industry will be forced to discard hierarchies, coercion, and external controls as the means of inducing compliance. Thus, technology may cure itself by simply evolving. The nagging doubt remains, however, whether those who control the technology will allow industry to evolve to this point. The experience of using automation, as I note in the final pages of chapter 2, does not seem to support Faunce's optimistic projections into the future. To date, industrial climates appear to have become more rather than less controlling with the advent of automation.

To summarize, the proposed solutions reviewed here appear unworkable simply because, as I mentioned earlier, they do not take into account psychological processes that are set in motion by the control of technology, nor the dynamic transformations in social behavior that occur as a result of this control. Perhaps the contribution of this book is to focus attention on those persons who use and control technology and the importance of understanding how the added power provided by technology transforms their behavior. While no solutions to the many problems of technology are offered in this book, I assume that the redefinition of the problem to include the issue of how the control of power changes people may provide a new perspective in the search for such solutions. To put the matter in the simplest of terms, knowledge may give us power, but power hardens the heart in predictable ways. As a consequence, while controllers of technology can accept as their due dominion over the earth, they can make little sense of the question: "Am I my brother's keeper?"

References

1. Plato (1955). *The republic.* (D. Lee, Trans.). England: Penguin Classics.
2. More, T. (1967). *Utopia and other essential writings.* New York: Meridian Classic.
3. Florman, S.C. (1981). *Blaming technology: The irrational search for scapegoats.* New York: St. Martin's Press.
4. Weinberg, A. (1966). Can technology replace social engineering? *University of Chicago Magazine, 59,* 6–10.
5. Dickson, D. (1974). *Alternative technology and the politics of technical change.* Glasgow: Fontana/Collins.
6. McDermott, J. (1969). Technology: The opiate of the intellectuals. *New York Review of Books, July 31, 1969.*
7. Jones, B. (1982). *Sleepers, wake! technology and the future of work.* England: Wheatsheaf Books LTD.
8. Cooley, M. (1980). *Architect or bee?* Boston, MA: South End Press.
9. Ellul, J. (1964). *The technological society.* New York: Alfred A. Knopf.
10. Juenger, F.G. (1949). *The failure of technology.* Chicago: Henry Regnery Co.
11. Mesthene, E.M. (1969). The role of technology in society. In A.H. Teich (Ed.), *Technology and man's future.* (pp. 99–129). New York: St. Martin's Press.
12. Franklin, U.M. (1984). Knowledge reconsidered: A feminist overview. *Proceedings of Annual Conference, Canadian Research Institute for the Advancement of Women.* Ottawa, Canada.
13. Mumford, L. (1967). *The myth of the machine.* London: Secker & Warburg.
14. Shaiken, H. (1985). *Work transformed.* New York: Holt Rinehart and Winston.
15. Baron, J.N., & Bielby, W.T. (1982). Workers and machines: Dimensions and determinants of technical relations in the workplace. *American Sociological Review, 47,* 175–188.
16. Easton, L., & Guddat, K.H. (1967). *Writings of the young Marx on philosophy and society.* Garden City. NY: Anchor Books.

17. Faunce, W.A. (1981). *Problems of an industrial society.* New York: McGraw-Hill.
18. Braverman, H. (1974). *Labor and monopoly capital.* New York: Monthly Review Press.
19. Reiman, B.C., & Inzerilli, G. (1978). *Technology and organization: A review and synthesis of major research findings.* Paper presented at the Conference on the Functioning of Complex Organizations, West Berlin, West Germany.
20. Edwards, R.C. (1979). *Contested terrain.* New York: Basic Books.
21. Lapp, R. (1965). *The new priesthood.* New York: Basic Books.
22. Hackman, R., & Oldham, G. (1980). *Work redesign.* Reading, MA: Addison-Wesley.
23. Schumacher, E.F. (1973). *Small is beautiful: Economics as if people mattered.* New York: Harper and Row.
24. Goodman, P. (1969). Can technology be humane? In C. Mitcham & R. Mackey (Eds.), *Philosophy and technology* (pp. 335–354). New York: St. Martin's Press.
25. Todd, J. (1977). A modest proposal. In N.J. Todd (Ed.), *The book of the new alchemists* (Chapter 1). New York: E.P. Dutton.
26. Coates, J.F. (1974). Technology Assessment. In McGraw Hill (Eds.), *Yearbook of science and technology* (pp. 65–74). New York: McGraw-Hill.
27. Gordon, T.R., & Ament, R.H. (1969). Forecasts of some technological and scientific developments and their societal consequences. In C. Mitcham & R. Mackey (Eds.), *Philosophy and Technology* (pp. 171–189). New York: St. Martin's Press.
28. Meadows, D., & Meadows, D.L. (1972). *The limits to growth.* New York: Universe Books.
29. Hellman, H. (1976). *Technophobia: getting out of the technology trap.* New York: M. Evans and Co.
30. Cornfield, D.B. (1987). *Workers, managers, and technological change: Emerging patterns of labor relations.* New York: Plenum Press.
31. Skinner, B.F. (1972). *Beyond freedom and dignity.* New York: Bantam Books.
32. Nelkin, D. (1980). Science and technology policy and the democratic process. In *The five year outlook: Problems, opportunities, and constraints in science and technology,* Vol. 2, (pp. 483–492). National Science Foundation, Washington, D.C.: Government Printing Office.
33. Mulder, M. (1971). Power equalization through participation. *Administrative science quarterly, March,* 31–38.
34. Winn, M. (1977). *The plug-in drug.* New York: Norton.

Author Index

Subject Index